Physics for the Logic Stage

Teacher Guide

Physics for the Logic Stage Teacher Guide

First Edition, 2nd Printing

Email: support@elementalscience.com

ISBN# 978-1-935614-40-1

Printed in the USA for worldwide distribution

For more copies write to:
Elemental Science
PO Box 79
Niceville, FL 32588
support@elementalscience.com

Copyright Policy

Physics for the Logic Stage
Teacher Guide Table of Contents

Physics for the Logic Stage
Introduction

In *Success in Science: A Manual for Excellence in Science Education*, we state that the middle school student is "a bucket full of unorganized information that needs to be filed away and stored in a cabinet."[1] The goals of science instruction at the logic level are to begin to train students' brain to think analytically about the facts of science, to familiarize the students with the basics of the scientific method through inquiry-based techniques, and to continue to feed the students with information about the world around them. *Physics for the Logic Stage* integrates the above goals using the Classic Method of middle school science instruction as suggested in our book. This method is loosely based on the ideas for classical science education that are laid out in *The Well-Trained Mind: A Guide to Classical Education at Home* by Jessie Wise and Susan Wise Bauer.

This guide includes the four basic components of middle school science instruction as explained in *Success in Science*.

1. **Hands-on Inquiry –** Middle school students need to see real-life science, to build their problem-solving skills and to practice using the basics of the scientific method. This can be done through experiments or nature studies. In this guide, the weekly experiments fulfill this section of middle school science instruction.
2. **Information –** Middle school students need to continue to build their knowledge base, along with learning how to organize and store the information they are studying. The information component is an integral part of this process. In this guide, the reading assignments, vocabulary, and sketches contain all of the necessary pieces of this aspect of middle school science instruction.
3. **Writing –** The purpose of the writing component is to teach students how to process and organize information. You want them to be able to read a passage, pull out the main ideas and communicate them to you in their own words. The assigned outlines or reports in this guide give you the tools you need to teach this basic component to your student.
4. **The Science Project –** Once a year, all middle school students should complete a science project. Their projects should work through the scientific method from start to finish on a basic level, meaning that their questions should be relatively easy to answer. The science fair project, scheduled as a part of unit three fulfills the requirements of this component.

Physics for the Logic Stage also includes the two optional components of middle school science instruction, as explained in *Success in Science.*

1. **Around the Web –** Middle school students should gain some experience with researching on the Internet. So for this optional component, the students should, under your supervision, search the Internet for websites, YouTube videos, virtual tours, and activities that relate to what they are studying. In this guide, the "Want More" lessons recommend specific sites and activities for you to use.
2. **Quizzes or Tests –** During the middle school years it is not absolutely necessary that you

[1]Bradley R. Hudson & Paige Hudson, *Success in Science: A Manual for Excellence in Science Education* (Elemental Science, 2012) 52.

give quizzes or tests to the students. However, if you want to familiarize them with test-taking skills, we suggest that you give quizzes or tests that will set the students up for success. With that in mind, we have included optional tests for you to use with each unit.

My goal in writing this curriculum is to provide you with the tools to explore the field of physics while teaching the basics of the scientific method. During the years, the students will work on their observation skills, learn to think critically about the information they are studying and practice working independently. *Physics for the Logic Stage* is intended to be used with seventh through eighth grade students.

What this guide contains in a nutshell

This guide includes the weekly student assignment sheets, all the sketches pre-labeled for you and discussion questions to help you guide your discussion time. This guide also contains information for each experiment, including the expected results and an explanation of those results. There is a list of additional activities that you can choose to assign for each week. Finally, this guide includes possible schedules for you to use as you guide the students through *Physics for the Logic Stage*.

What the Student Guide contains

The Student Guide, which is sold separately, is designed to encourage independence in the students as they complete *Physics for the Logic Stage*. The Student Guide contains all the student assignment sheets, pre-drawn sketches ready for labeling, experiment pages, and blank report pages. The guide also includes blank date sheets as well as all the sheets they will need for the Science Fair Project. In short, the Student Guide contains all the pages the students will need and it is essential for successfully completing this program.

Student Assignment Sheets

This Teacher Guide contains a copy of each of the student assignment sheets that are in the Student Guide. This way you can stay on top of what your students are studying. Each of the student assignment sheets contains the following:

✓ **Experiment**

Each week will revolve around a weekly topic that it to be studied. Your student will be assigned an experiment that poses a question related to the topic. Each of these experiments will walk your students through the scientific method. (*See the Appendix pg. 245 for a brief explanation of the scientific method.*) In a nutshell, the scientific method trains the brain to examine and observe before making a statement of fact. It will teach your student to look at all the facts and results before drawing a conclusion. If this sounds intimidating, it's not. You are simply teaching your students to take the time to discover the answer to a given problem by using the knowledge they have and the things they observe during an experiment.

Each week, the student assignment sheet will contain a list of the materials needed and the instructions to complete the experiment. The student guide contains an experiment sheet for your students to fill out. Each experiment sheet contains an introduction that is followed by a list of materials, a hypothesis, a procedure, an observation, and a conclusion section. The introduction will give your students specific background information for the experiment. In the hypothesis section, they will predict the answer to the question posed in the lab. In the materials listed section, your students will fill out what they will use to complete the experiment. In the procedure section, they will recount step by step what was done during their experiment, so that someone else could read their report and replicate their experiment. In the observation section, your students will write what they saw. Finally, in the conclusion section they will write whether or not their hypothesis was correct and share any additional information they have learned from the experiment. If the students' hypotheses were not correct, discuss why and have them include that on their experiment sheet.

Vocabulary & Memory Work

Throughout the year, the students will be assigned vocabulary for each week. They will need to write out the definitions for each word on the Unit Vocabulary Sheet found in the Student Guide on the week that they are assigned. You may want to have your students also make flash cards to help them work on memorizing the words. This year, the students will memorize the elements of the periodic table along with specific information relating to each unit. There is a complete listing of the vocabulary words and memory work for each unit on the unit overview sheet in this guide along with a glossary and a list of the memory work in the Student Guide.

Sketch

Each week the students will be assigned a sketch to complete and label. The Student Guide contains an unlabeled sketch for them to use. They will color the sketch, label it and give it a title according to the directions on the Student Assignment Sheet. The information they need will be in their reading, but the sketch is not always identical to the pictures found in the encyclopedia. So, these sketch assignments should make the student think. This guide contains a completed sketch for you to use when checking their work.

Writing

Each week the students will be assigned pages to read from the spine text, the *DK Encyclopedia of Science*. Have them read the assigned pages and discuss what they have read with you. After you have finished reading and discussing the information, you have three options for your students' written assignments:

- ***Option 1: Have the students write an outline from the spine text***
  A typical seventh grader completing this program should be expected to write a two to three level outline for the pages assigned for the week. This outline should include the main point from each paragraph on the page as well as several supporting and sub supporting points;
- ***Option 2: Have the students write a narrative summary from the spine text***
  A typical seventh grader completing this program should be expected to write a three to six paragraph summary (or about a page) about what they have read in the spine text;
- ***Option 3: Have the students write both an outline and a written report***
  First, have the students read the assigned pages in the spine text. Then, have them write a two to three level outline for the assigned pages. Next, have the students do

some additional research reading on the topic from one or more of the suggested reference books listed below. Each topic will have pages assigned from these reference books for their research. The following encyclopedias are scheduled to be used as reference books:

- *The Kingfisher Science Encyclopedia, 2017 Edition (KSE):* This resource is appropriate for middle school students.
- *Usborne Illustrated Dictionary of Science, 2012 Edition (UIDS):* This resource is approaching the high school level.

Once the students complete the additional research reading, have them write a report of three to four paragraphs in length, detailing what they have learned from their research reading.

Your writing goal for middle school students is to have them write something (narrative summary, outline, or list of facts) every day you do school, either in science or in another subject. So, the writing option you choose for this curriculum will depend on the writing the students are already doing in their other subjects.

When evaluating the students' reports, make sure that the information they have shared is accurate and that it has been presented in a grammatically correct form (i.e., look for spelling mistakes, run-on sentences, and paragraph form). In the Student Guide, there are two blank lined sheets for the students to use when writing their outlines and/or summaries. If you are having the students type their report, have them glue a copy of it into their Student Guide.

Dates

Each week the dates of important discoveries within the topic and the dates from the readings are given on the student assignment sheet. The students will enter these dates onto one of their date sheets. The date sheets are divided into the four time periods as laid out in *The Well Trained Mind* by Susan Wise Bauer and Jessie Wise (Ancients, Medieval-Early Renaissance, Late Renaissance-Early Modern, and Modern). Completed date sheets are available for you to use in the appendix of this guide on pg. 241-244.

Schedules

Physics for the Logic Stage is designed to take up to 5 hours per week. You and your students can choose whether to complete the work over five days or over two days. Each week I have included two scheduling options for you to use as you lead them through this program. They are meant to be guides, so feel free to change the order to better fit the needs of your students. I also recommend that you begin to let them be in charge of choosing how many days they would like to do science as this will help to begin to foster independence in their school work.

Additional Information Section

The Additional Information Section includes tools that you will find helpful as you guide the students through this study. It is only found in the Teacher Guide, and it contains the following:

Experiment Information

Each week, the Additional Information Section includes the expected experiment results and an explanation of those results for you to use with the students. When possible, you will also find suggestions on how to expand the experiment in the Take if Further section.

Discussion Questions

Each week the Additional Information Section includes possible discussion questions from the main reading assignment, along with the answers. These are designed to aid you in leading the discussion time with the students. I recommend that you encourage them to answer in complete sentences, as this will help them organize their thoughts for writing their outline or report. If the students are already writing outlines or lists of facts, you do not need to have them write out the answers to the discussion questions before hand as there is plenty of writing required in this program already.

Want More

Each week, the Additional Information Section includes a list of activities under the Want More section. ***These activities are totally optional.*** The Want More activities are designed to explore the science on a deeper level by researching specific topics or through additional projects to do. The students do not have this information in their guide, so it is up to you whether or not to assign these.

☑ **Sketch**

Each week, the Additional Information Section includes copies of the sketches that have been labeled. These are included in this guide for you to use as you correct the students' work.

Tests

The students will be completing a lot of work each week that will help you to assess what they are learning, so testing is not absolutely necessary. However, I have included end of unit tests that you can use if you feel the need to do so. The tests and the answers are included after the material for each unit in this guide. You can choose to give the tests orally or copy them for the students to fill out.

What a typical two day schedule looks like

A typical two day schedule will take one and a half to two hours per day. Here's a breakdown of how a normal two day week would work using week two:

- **Day 1: *Define the vocabulary, record the dates, do the experiment, and complete the experiment sheet***
  Begin day 1 by having the students do the "How does friction affect movement?" experiment. Have them read the introduction and perform the experiment using the directions provided. Next, have the students record their observations and results. After they discuss their results with you, have them write a conclusion for their experiment. Finish the day by having them look up and define "air resistance," "friction," "gravity," and "terminal velocity" using the glossary in the Student Guide and add the dates to their date sheets.
- **Day 2: *Read the assigned pages, discuss together, prepare an outline or narrative summary, and complete the sketch***
  Begin by having the students read pp. 121-122 in the *DK Encyclopedia of Science*. Then, using the questions provided, discuss what they have read. Next, have them

complete the sketch using the directions on the Student Assignment Sheet. Finally, have them write an outline or narrative summary. Here is a sample narrative summary:

Friction and Gravity

Friction is the force responsible for slowing down the movement of objects as they slide over each other. The rougher the surface or the heavier an object, the more friction is produced.

Without friction, we could not do a lot of things. For example, we could not walk because our shoes could not grip the ground without friction, and we could not grip because our fingers would not be able to hold the object. Friction is the force responsible for slowing a vehicle down when we hit the brakes.

Friction causes wear and tear on machines, but we can reduce friction an object experiences. We can use lubricants, like oil, reduce the friction. We can also use ball bearings, which reduce friction by causing objects to roll over each other instead of dragging.

Gravity is the force that pulls two objects together. The force of this gravitational pull is dependent upon the distance between the two objects and the mass of the two objects. The closer the objects are, the greater the force of gravity between them. Objects that have a great mass produce a larger gravitational force.

Gravity is the force responsible for creating the tides in the ocean. The gravity of the moon pulls on the ocean on the side of the Earth that is closest, causing it to bulge out. As the two bodies rotate around each other, the strength of the force changes, causing the tides.

What a typical five day schedule looks like

A typical five day schedule will take forty-five minutes to one hour per day. Here's a breakdown of how a normal five day week would work using week two:

- **Day 1:** ***Do the experiment and complete the experiment sheet***
  Begin day 1 by having the students do the "How does friction affect movement?" experiment. Have them read the introduction and perform the experiment using the directions provided. Next, have them record their observations and results, discuss their results with you, and then write a conclusion for their experiment.
- **Day 2:** ***Read the assigned pages, discuss together and write an outline or list of facts***
  Begin by having the students read pp. 121-122 in the *DK Encyclopedia of Science* and discuss what they have read using the provided questions. Then, have the students write a two to three level outline, and complete the sketch using the directions on the Student Assignment Sheet. Here's a sample outline for the page on friction:

Friction

I. Force which slows down the movement of objects as they slide over each other.
   A. The rougher the surface, the more friction there is.
   B. Heavy objects would be easy to move without friction.
II. Without friction, we could not do a lot of things.

A. We could not walk because our shoes could not grip the ground without friction.
B. We could not grip because our fingers would not be able to hold the object.

III. Friction causes wear and tear on machines.
A. You can reduce friction.
i. Lubricants, like oil, reduce friction.
ii. Ball bearings reduce friction because they cause objects to roll over each other instead of dragging.

IV. Friction is everywhere.
A. There is friction between the brake pads and wheels on a bike.
B. There is friction between gears.
C. There is friction as an object moves through water.

V. Friction in the air.
A. Air resistance is the friction force that objects feel as they move through the air.
B. The faster an object is moving, the more air resistance it feels.
C. Friction can heat things up, which is why a meteor burns up as it travels through our atmosphere.

- **Day 3: *Record the dates, define the vocabulary, and complete the sketch***
Begin by having the students look up and define "air resistance," "friction," "gravity," and "terminal velocity" using the glossary in the Student Guide and add the dates to their date sheets. Then, have them complete the sketch using the directions on the Student Assignment Sheet.
- **Day 4: *Read from the additional reading assignments and prepare a written report***
Begin by having the students read "Relativity and Gravity" from *KSE* pp. 298-299, "Friction" from *KSE* pp. 308-309, or "Gravitation" from *UDIS* pp. 18-19. Then, have the students use their outline along with what they have just read to write a three to five paragraph summary of what they have learned. Here is a sample report:

Gravity, Weight, and Realtivity

There is an attractive force that exists between all masses, which is known as gravity. The strength of the force between two objects depends upon the distance between them and their masses. Mass is a measure of the amount of matter in an object, which should not be confused with weight. Weight is the force experience by a given amount of mater within a gravitational field.

On Earth, the force of the gravity we feel would cause use to accelerate at 9.8 meters per second. The force of gravity on other planets is larger or smaller depending on the planet's size. However out in space, we are weightless because there is no gravitational force pulling on us.

Albert Einstein published a theory that showed nothing could travel faster than the speed of light. This theory conflicted with the idea *that there must be a gravitational pull to be able to travel at infinite speed, which was presented by Isaac Newton. Einstein fixed this with his General Theory of Relativity that describes gravity as a distortion of space and time.*

- **Day 5: *Complete one of the Want More activities***
  Have the students do the "Galileo's Tracks" activities or have them do the "Friction Demonstrion" on-line. You could also have them read about a scientist from the field of physics.

The Science Fair Project

I have scheduled time for the students to complete a science fair project during unit three. Janice VanCleave's *A+ Science Fair Projects* & Janice VanCleave's *A+ Projects in Physics: Winning Experiments for Science Fairs and Extra Credit* are excellent resources for choosing project topics within the field of physics. You can call your local school system to see if it allows homeschooled students to participate in the local school science fair or get information on national science fairs from them. Another option would be to have your students present their project in front of a group of friends and family.

How to include your younger students

I recognize that many homeschool families have a range of different student ages. If you wish to have all your students studying the topic of physics you have two options for your elementary students when using this program with your middle school students:

- ***Option 1: Have your younger students use Physics for the Grammar Stage***
  I recommend this option if your younger students are in the second through fourth grade and/or your older students are ready for some independence. You will need to rearrange the units in *Physics for the Logic Stage* so that all the students will remain on similar topics. The older students will do Unit 2, Units 1, and then Unit 3 through Unit 8.

- ***Option 2: Have your younger students use Physics for the Logic Stage along with your older students***
  I recommend this option if your younger students are in the fourth through sixth grade and/or older students are not ready to work independently. However, you will need to adjust the work load for your younger students. Here are some suggestions on how to do that:
  - ✓ Have them watch and observe the experiments;
  - ✓ Add in some picture books from the library for each of the topics;
  - ✓ Read the reading assignments to them and have them narrate them back to you;
  - ✓ Let them color the sketches and then tell them how to label them.

  As for the reading assignments, you may find that the spines scheduled are too much for your younger students. If so, you can read to them out of the *Usborne Science Encyclopedia*. I have included a chart coordinating this resource in the Appendix of this guide on pg. 247-249.

Helpful Articles

Our goal is to provide you with the information you need to be successful in your quest to educate your students in the sciences at home. This is the main reason we share tips and tools for homeschool science education at our blogs. As you prepare to guide your students through this program, you may find the following articles from there helpful:

- *Classical Science Curriculum for the Logic Stage Student* – This article explains the goals of logic stage science and demonstrates how the classical educator can utilize the tools they have at their disposal to reach these goals.
  - http://elementalblogging.com/classical-science-curriculum-logic/
- *Scientific Demonstrations vs. Experiments* – This article shares information about these two types of scientific tests and points out how to employ scientific demonstrations or experiments in your homeschool.
  - https://elementalscience.com/blogs/news/89905795-scientific-demonstrations-or-experiments
- *A Simple Explanation of the Scientific Method* – This article details the steps of the scientific method, along with why it is so important to teach.
  - https://elementalscience.com/blogs/news/simple-explanation-of-the-scientific-method/

Additional Resources

The following page contains quick links to the activities suggested in this guide along with several helpful downloads:

- https://elementalscience.com/blogs/resources/pls

Final Thoughts

If you find that this program contains too much work, please tailor it to the needs of your students. As the author and publisher of this curriculum I encourage you to contact me with any questions or problems that you might have concerning *Physics for the Logic Stage* at support@ elementalscience.com. I will be more than happy to answer them as soon as I am able. I hope that you and your students enjoy *Physics for the Logic Stage*!

Book List

The following books were used when planning this study:

Encyclopedias for Reading Assignments

The following book is the main spine of this program. You will need to purchase both of these to complete the reading assignments scheduled in this program.

- 📖 *The DK Encyclopedia of Science, 2016 Edition (DK EOS)*
- 📖 *Bridges and Tunnels* by Donna Latham
- 📖 *Robotics* by Kathy Ceceri

References for Reports

The following encyclopedias are scheduled for additional reference reading. They are optional, but I suggest that you purchase one or two to use throughout the year.

- 📖 *The Kingfisher Science Encyclopedia, 2017 Edition (KSE)* – This resource is appropriate for middle school students.
- 📖 *Usborne Illustrated Dictionary of Science, 2012 Edition (UIDS)* – This resource is approaching the high school level.

Experiment Equipment

If you would like to create a more lab-like experience for the students this year, I suggest using equipment that is more commonly found in the laboratory setting. Here's a list of material that you can substitute:

- ✓ **Jar** – Use a beaker or Erlenmeyer flask that is at between 750 and 1000 mL;
- ✓ **Cup** – Use a beaker or Erlenmeyer flask that is at between 200 and 500 mL;
- ✓ **Bottle** – Use an Erlenmeyer flask that is between 250 and 1000 mL;
- ✓ **Small cup** – Use a small beaker (50 mL) or test tube;
- ✓ **Eye dropper** – Use a pipette.

You can use the glass or plastic version of each of the above.

Safety Advisory

Some of the experiments in this book use boiling water or open flames. We recommend that your students use safety glasses and protective gear with each experiment to prevent accidents. Do not allow your students to perform any of the experiments marked " ☢ **CAUTION** " on their own.

Units of Measurement

What are the two measuring systems?

- **The Standard or Standard American Engineering (SAE) System –** This system is used mainly in the United States and it uses units like inches, pounds and gallons. It was derived from an early English measuring system that has its roots in the Roman system of measurements.
- **The Metric System –** This system is used in most of the world and it uses units like meters, grams and liters. The system is base 10 and their names are formed with prefixes. It was derived from one of the early French measuring systems.

In the US, the standard system of units are more widely used on consumer products and in industrial manufacturing, while the metric system is more widely used in science, medicine and government. Since this program has been published in the US, I have used the standard system of measurement throughout for familiarity. However, because I believe that it is important for our students to be familiar with both systems, I have included metric measurements in parentheses.

What about converting units?

Every student should know how to convert measurements inside of a given measuring system, such as knowing how to convert grams to kilograms or ounces to pounds. Normally, these conversion factors are taught as a part of your math program. However, I also recommend that you have your students memorize several basic conversion factors between the two systems. Here is a list of factors that the students should try to memorize:

- **Pounds to Kilograms:** 1 lb = 2.2 kg
- **Ounces to Grams:** 1 oz = 28.3 g
- **Gallons to Liters:** 1 gal = 3.785 L
- **Cups to Milliliters:** 1 c = 240 mL
- **Miles to Kilometers:** 1 mi = 1.61 km
- **Feet to Meters:** 1 ft = 0.305 m
- **Inches to Centimeters:** 1 in = 2.54 cm

With the global flow of information that occurs these days, it is very important for students to learn these most basic conversion factors. To learn more about the importance of units of measurement in science, read the following blog post:

- https://elementalscience.com/blogs/science-activities/units-of-measurement

Sequence of Study

Basics of Physics - Forces, Motion , and Energy

Unit 1: Motion (4 Weeks)

- ✓ Forces
- ✓ Friction and Gravity
- ✓ Motion
- ✓ Speed and Acceleration

Unit 2: Energy (4 Weeks)

- ✓ Energy and Work
- ✓ Energy Sources
- ✓ Pressure
- ✓ Simple Machines

Concepts in Physics - Heat, Light, and Sound

Unit 3: Thermodynamics (4 weeks)

- ✓ Energy Conversion
- ✓ Heat
- ✓ Thermodynamics
- ✓ Engines
- ✓ Science Fair Project

Unit 4: Sound (4 Weeks)

- ✓ Sound
- ✓ Sound Waves
- ✓ Hearing Sound
- ✓ Acoustics

Unit 5: Light (4 Weeks)

- ✓ Light
- ✓ Reflection and Refraction
- ✓ Vision and Color
- ✓ Optics

Applications in Physics - Electricity, Magnetism, and Engineering

Unit 6: Electricity and Magnetism (7 Weeks)

- ✓ Electrical Current
- ✓ Conductors and Insulators
- ✓ Batteries
- ✓ Circuits
- ✓ Magnetism

- ✓ Electromagnetism
- ✓ Motors and Generators

Unit 7: Engineering and Robotics (6 Weeks)

- ✓ Engineering
- ✓ Bridges
- ✓ Tunnels
- ✓ Robotics
- ✓ Actuators and Effectors
- ✓ Sensors and Controllers

Unit 8: Nuclear Physics (2 weeks)

- ✓ Radioactivity
- ✓ Nuclear energy
- ✓ Scientist Study

Year-end Review

Review Test

Materials Listed by Week

Basics of Physics - Forces, Motion, and Energy

Unit 1: Motion

Week	Materials
1	Thick, sturdy cardboard, 1 Brad fastener, Rubber band, Hole punch or nail, String – about 3 in (10 cm), 3 Jumbo paper clips, Pen, Objects of varying weight
2	Force meter from last week, Small wooden block (like a Jenga block), Eye-hook screw, Sandpaper, Felt, Foil, Spray oil, Tape measure
3	Jenga block with the eyehook from last week, String, 2 Toy cars, Egg
4	Cardboard or plastic track, Blocks or thick books, Toy car, Stopwatch

Unit 2: Energy

Week	Materials
5	Goldfish cracker, Small marshmallow, Piece of lettuce, Piece of bacon fat, Aluminum pan, Matches, Safety glasses, Bucket of water
6	2-Liter Soda bottle, 2 Cans – one large, one small, Screw, Water, Piece of clay, Cup measure, Tape measure
7	Foil, Black construction paper, Small cardboard box, Plastic wrap, Tape, Marshmallow, Small glass dish (one that will fit inside the box)
8	Materials will vary depending upon the simple machine the student chooses to build

Concepts in Physics - Heat, Light, and Sound

Unit 3: Thermodynamics

Week	Materials
9-12	*Science Fair Project supplies will vary depending on the project the students choose to do.*

Unit 4: Sound

Week	Materials
13	Glass bottle, Bell, Cork that fits the top of the glass bottle, Thread, Needle, Match
14	Shallow glass bowl or cup, Water, Music player
15	Plastic jar, or small flower pot, A piece of latex material large enough to cover the lid of your jar (like the kind used for exercise bands), 1" plastic tubing, Rubber band, Air-dry clay, Salt
16	Partner, Blindfold

Unit 5: Light

Week	Materials
17	9 Ultraviolet light detecting beads, 3 Shallow dishes (not clear plastic or glass), Plastic Wrap, Two different levels of SPF sunscreen (i.e., SPF 15 and SPF 45), Rubber bands
18	4 Pencils, 4 Clear glasses, Water, Oil, Alcohol, Corn syrup
19	Thin cardboard, Red, blue, and yellow paint, 6 Rubber bands, Hole punch
20	Jell-O™ (orange, lemon, or lime), Round bowl or jar – at least 4" (10 cm) in diameter, 1 Cup water, Dull knife, Plate, Flashlight

Applications in Physics - Electricity, Magnetism, and Engineering

Unit 6: Electricity and Magnestism

Week	Materials
21	Styrofoam pan, Aluminum pan, Wool, Plastic tongs
22	Light bulb, Copper wire, D battery, Electrical tape, Alligator clips, Organic material, such as a pickle, lemon slice, cheese, bread, or leaf
23	2 AA disposable batteries (one fully charged, one completely dead), Ruler
24	Computer with Internet connection
25	2 different types of magnets, such as a horseshoe magnet and a neodymium magnet, Paper clips (20 to 30), Paper, Cardboard, Thick books
26	D battery, Insulated copper wire – about 3 ft (1 m), 2 to 3 inch (5 to 8 cm) Nail, Electrical tape, Iron filings, Paper
27	Straws, Electrical tape, 6 ft. (2 m) of thin insulated wire, AAA battery, Sandpaper, Needle

Unit 7: Engineering

Week	Materials
28	Paper, Tape, Books, Can or glass
29	Craft sticks, Wood glue, Books, Binder clips
30	Salt dough (at least 3 to 4 cups), Cardboard square, Spoon, Craft sticks, Pipe cleaners, Aluminum foil, Toy car, Books or other heavy objects, Water
31	1.5-volt DC motor, 1 ft. insulated wire, Electrical tape, Cup or Jar, Foam tape, 2 AAA batteries, Rubber band, Cork, Cardboard, 3 Pens, Paper
32	Pencil, 1.5-volt DC motor, Small Solar Panel, Electrical tape, Scissors, CD, Glue, Tape, Clear dome from a drink cup
33	LED light bulb with two metal legs, 3-volt Watch battery, 2 Index cards, Aluminum foil, Scissors, Marker, Yarn, Glue, Toothpick, Tissue

Unit 8: Nuclear Physics - No supplies needed.

Physics Unit 1

Forces and Motion

Unit 1 Force and Motion
Overview of Study

Sequence of Study

Week 1: Force
Week 2: Friction and Gravity
Week 3: Motion
Week 4: Speed and Acceleration

Materials by Week

Week	Materials
1	Thick, sturdy cardboard, 1 Brad fastener, Rubber band, Hole punch or nail, String – about 3 in (10 cm), 3 Jumbo paper clips, Pen, Objects of varying weight
2	Force meter from last week, Small wooden block (like a Jenga block), Eye-hook screw, Sandpaper, Felt, Foil, Spray oil, Tape measure
3	Jenga block with the eyehook from last week, String, 2 Toy cars, Egg
4	Cardboard or plastic track, Blocks or thick books, Toy car, Stopwatch

Vocabulary for the Unit

1. **Balance** – A state of equilibrium when the forces acting on an object cancel each other out ; also known as a zero resultant force.
2. **Force** – A push or pull that acts on an object.
3. **Force field** – The area in which a force can be felt.
4. **Newton** – The measurement of force; one newton is the force is takes to move a one kilogram object at one meter per second squared ($1 N = 1 kg \cdot 1 m/s^2$).
5. **Air resistance –** The force that air exerts on an object as it falls.
6. **Friction –** A force that opposes the motion of objects that touch as they move past each other.
7. **Gravity** – The force that acts between two masses; it is an attractive force.
8. **Terminal velocity** – The point at which the force acting on an object of air resistance is equal to the force of gravity acting on the object.
9. **Inertia –** The tendency of an object to resist a change in its motion.
10. **Mass –** The amount of matter in an object.
11. **Momentum –** The tendency of an object to keep moving until a force stops it.
12. **Weight –** The force with which an object's mass is pulled toward the center of the Earth.
13. **Acceleration –** A change in an object's speed, direction, or both.

14. Speed – The ratio of the distance an object moves to the amount of time the object moves.
15. Velocity – The speed of an object in a particular direction.

Memory Work for the Unit

Newton's Three Laws of Motion

1. An object will not move unless a force, like a push or pull, moves it. Once it is moving, an object will not stop moving in a straight line unless it's forced to change.
2. The greater the force on an object, the greater the change in its motion. The greater the mass of an object, the greater the force needed to change its motion.
3. For every reaction, there is an equal but opposite reaction.

Equations

- Force Unit

  1 Newton (N) = 1 kilogram (kg) • 1 meter (m) / second (s^2)

- Motion Equation

  $F = m \cdot A$

  "F" stands for net force.
  "m" stands for mass.
  "A" stands for acceleration.

- Speed Equation

  $v = \frac{d}{t}$

  "v" stands for average speed.
  "d" stands for distance.
  "t" stands for time.

- Acceleration Equation

  $A = \frac{v_f - v_i}{t}$

  "A" stands for acceleration.
  "v_f" stands for final speed.
  "v_i" stands for initial speed.
  "t" stands for time.

Notes

Student Assignment Sheet Week 1
Forces

Experiment: Can I Measure Force?

Materials

- ✓ Thick, sturdy cardboard
- ✓ 1 Brad fastener
- ✓ Rubber band
- ✓ Hole punch or nail
- ✓ String – about 3 in (10 cm)
- ✓ 3 Jumbo paper clips
- ✓ Pen
- ✓ Objects of varying weight

Procedure

1. Read the introduction to the experiment and then begin to assemble your force meter. Cut out a 3.5 in (9 cm) by 12 in (31 cm) rectangle from the cardboard. Then, punch a hole with the hole punch or nail near the top, large enough for the brad fastener to slide through. Slip one of the paper clips through the brad, through the hole, and fasten the brad on the opposite side. Slide the rubber band onto the opposite end of the paper clip. Next, take another paper clip and turn out a portion of the end to make a pointer. Tie the string to one end of the pointer paper clip and then slide the other end onto the rubber band. Take the third paper clip and fashion a hook out of it. Once you are done, attach the hook to the other end of the string. Hold your force meter at the top and mark where the pointer rests. This line will be your zero force mark. Now draw a scale down the remainder of your force meter. You can use finger widths, inches, or centimeters for your scale, just as long as you use the same measurement for each mark. (**Note**—*You will need your force meter for next week's experiment as well.*)
2. Now that the force meter is assembled, you can use it to measure the different objects. Simply attach each object to the hook and observe what happens. Write down how much the rubber band stretched on the experiment sheet. Repeat this process for each of your objects.
3. Draw conclusions and complete the experiment sheet.

Vocabulary & Memory Work

- ☐ Vocabulary: balance, force, force field, newton
- ☐ Memory Work—This week, work on memorizing the force equation:
  - 1 Newton (N) = 1 kilogram (kg) • 1 meter (m) / second (s^2)

Sketch: Resultant Force

- Label the following—Forces pull in the same direction; add the forces together to get the resultant force; forces pull in equal, but opposite directions; the forces will cancel each other out for a zero resultant force; forces pull unequal, opposite directions; subtract the forces to get the resultant force.

Writing

- Reading Assignment: *DK Encyclopedia of Science* pp. 114-115 (Forces), pg. 116 (Combining Forces), and pg. 117 (Balanced Forces)
- Additional Research Readings
  - Force: *KSE* pp. 290-291, *UDIS* pp. 6-7

Dates

- c330 BC – Aristotle proposes that a force is needed to maintain motion.
- 1642-1727 – Isaac Newton, the English scientist who explained how force, mass, and acceleration are related, lives. The unit of force, the newton (N), is named after him.
- 1979 – Pakistani scientist, Abdus Salam, wins the Nobel Prize in Physics for his work with forces.

Schedules for Week 1

Two Days a Week

Day 1	Day 2
☐ Do the "Can I Measure Force?" experiment, and then fill out the experiment sheet on SG pp. 20-21 ☐ Define balance, force, force field, and newton on SG pg. 16 ☐ Enter the dates onto the date sheets on SG pp. 8-13	☐ Read pp. 114-117 from *DK EOS,* and then discuss what was read ☐ Color and label the "Resultant Force" sketch on SG pg. 19 ☐ Prepare an outline or narrative summary; write it on SG pp. 22-23

Supplies I Need for the Week

✓ Thick, sturdy cardboard, 1 Brad fastener, Rubber band
✓ Hole punch or nail, String – about 3 in (10 cm)
✓ 3 Jumbo paper clips
✓ Pen, Objects of varying weight

Things I Need to Prepare

Five Days a Week

Day 1	Day 2	Day 3	Day 4	Day 5
☐ Do the "Can I Measure Force?" experiment, and then fill out the experiment sheet on SG pp. 20-21 ☐ Enter the dates onto the date sheets on SG pp. 8-13	☐ Read pp. 114-117 from *DK EOS,* and then discuss what was read ☐ Write an outline on SG pg. 22	☐ Define balance, force, force field, and newton on SG pg. 16 ☐ Color and label the "Resultant Force" sketch on SG pg. 19	☐ Read one or all of the additional reading assignments ☐ Write a report on what you learned on SG pg. 23	☐ Complete one of the Want More Activities listed **OR** ☐ Study a scientist from the field of Physics

Supplies I Need for the Week

✓ Thick, sturdy cardboard, 1 Brad fastener, Rubber band
✓ Hole punch or nail, String – about 3 in (10 cm)
✓ 3 Jumbo paper clips
✓ Pen, Objects of varying weight

Things I Need to Prepare

Additional Information Week 1

Notes

- **Mass vs. Weight** – Mass is the measurement of how much matter an object contains, whereas weight is the measurement of the pull of gravity on an object. The more mass an object contains, the more it weighs because there is more substance on which gravity can pull.

Experiment Information

- **Note** – Make sure your students keep their force meter for next week.
- **Introduction** – (*from the Student Guide*) Forces are all around us. They push and pull objects, causing them to move or change shape. In today's experiment, you are going to create your own force meter that can measure the amount of force an object exerts. In a force meter, an object applies a downward force, which stretches a rubber band or spring. We can measure the amount of displacement to determine how much force was applied.
- **Results** – The students' results will vary based on the objects that they choose to use. In general, they should see that a heavier object will cause the rubber band to stretch farther.
- **Explanation** – The weight of each objects acts as a force that pulls down on the rubber band, causing it to stretch. The more the object weighs, the greater the force, which causes the rubber band to stretch farther.
- **Troubleshooting Tips** – Be sure that the students use thick, sturdy cardboard when making their force meter or it can tear. To see a visual representation of a homemade force meter and how to calibrate it, check out the following video:
  - https://www.youtube.com/watch?v=jwCwwKLa0GE

  If they want to make a sturdier version out of PVC pipe, have them follow the directions from this website:
  - http://www.instructables.com/id/Be-a-scientist%3A-make-your-own-force-meter./
- **Take it Further** – Have the students read *DK Encyclopedia of Science* pg. 123 (Measuring Forces). Then, have them calibrate the force meter to actual newtons (N). One newton exerts approximately a quarter of a pound of force (or about 100 grams). Hang something that weighs 0.25 lbs (100 g) on your force scale and mark where the guide line lands – this is the 1N mark. Now, repeat the process up with objects weighing up to 1 lb to find where 2N, 3N and 4N would be. (**Note**—*Since the rubber band doesn't stretch linearly, the marks may not be evenly spaced.*)

Discussion Questions

Forces, pp. 114-115

1. What does a force do? (*A force acts on an object or a force pushes or pulls an object.*) Name several examples. (*The wind blowing, gravity pulling, and the grasshopper leaping are all examples of force.*)
2. Where is a force field the strongest? (*A force field is strongest closest to the source of the force.*)
3. What is the difference between contact and non-contact forces? (*Contact forces are only produced when one object touches another. Non-contact forces can pull objects without*

touching them.)

Combining Forces, pg. 116

1. What is a resultant force? (*The resultant force is the overall result of two or more forces acting on an object.*)
2. How do you find the resultant when forces are pulling in the same direction? (*When forces are pulling in the same direction, you can find the resultant by adding the forces together.*)
3. How do you find the resultant when forces are pulling in the opposite direction? (*When forces are pulling in the opposite direction, you can find the resultant by subtracting one force from the other.*)

Balanced Forces pg. 117

1. How is an object balanced? (*An object is balanced when the forces acting on it cancel each other out, which produces a zero resultant.*)
2. Why is balance important to architects? (*Architects design buildings and bridges so that the forces that act on the structure are balanced. This keeps the structure from falling down.*)

Want More

- **Tug of War** – In a tug of war, each team is using force to pull the other team across the line. One team's pulling force cancels out the other team's pulling force, which keeps the players at a stand-still. That is, until one team's pulling force is greater than the other's! This week, explain to your students how force plays a role in tug of war and then let them try it out for themselves. If you can't get a team together, have the students do the tug of war simulation from the PhET website.
  - http://phet.colorado.edu/sims/html/forces-and-motion-basics/latest/forces-and-motion-basics_en.html
- **Resultant Force Worksheet** – Have the students complete the resultant force worksheet on Appendix pg. 250. Answers
  1. Resultant force = 0N, object is in balance
  2. Resultant force = -2N, object will begin moving in the opposite direction
  3. Resultant force = 8N, object will continue in the same direction
  4. Resultant force = 0N, object is in balance

Sketch Week 1

Resultant Force

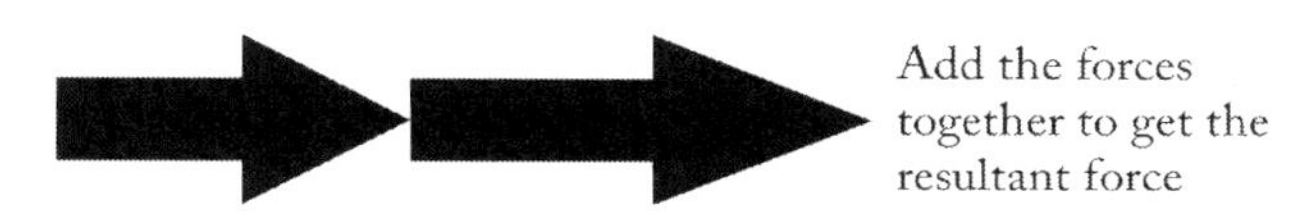

Add the forces together to get the resultant force

Forces pull in the same direction

Forces pull in equal, but opposite directions

The forces will cancel each other out for a zero resultant force

Forces pull unequal, opposite directions

Subtract the forces to get the resultant force

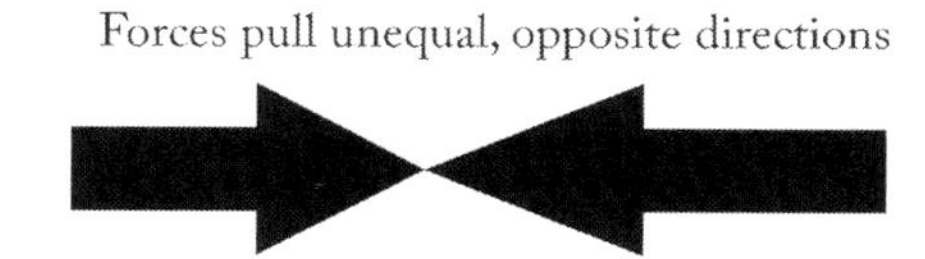

Student Assignment Sheet Week 2
Friction and Gravity

Experiment: How does friction affect movement?

Materials
- ✓ Force Meter from last week
- ✓ Small wooden block (aka. Jenga block)
- ✓ Eye-hook screw
- ✓ Sandpaper
- ✓ Felt
- ✓ Foil
- ✓ Spray oil
- ✓ Tape measure

Procedure
1. Read the introduction to the experiment and answer the question for the hypothesis section.
2. Screw the eye-hook screw into the top of the wooden block. Then, attach it to the hook on the force meter so that the block can be dragged horizontally. Next, use the tape measure to mark off a 1 foot (0.3 meter) track on a smooth surface, like a table our counter.
3. Now, place the block at the beginning of your track with the force meter in front over the track. Pull the block from the force meter evenly to the end in three seconds. Observe how much the rubber band on the force meter stretched and record that on your experiment sheet.
4. Then, place the piece of sandpaper on your track. Like before, put block at the beginning of the track and pull it evenly to the end in three seconds. Observe how much the rubber band on the force meter stretched and record that on your experiment sheet. Repeat with the felt.
5. Finally, place the foil over the track and coat it well with spray oil. Then, as before, put block at the beginning of your track and pull it evenly to the end in three seconds. Observe how much the rubber band on the force meter stretched and record that on your experiment sheet.
6. Draw conclusions and complete the experiment sheet.

Vocabulary & Memory Work
- ☐ Vocabulary: air resistance, friction, gravity, terminal velocity
- ☐ Memory Work—This week, begin working on memorizing Newton's three laws of motion. (*See Unit Overview Sheet for a complete listing.*)

Sketch: Types of Friction (*See the Sketch Notes.*)
- Label the following – Static friction, sliding friction, rolling friction, fluid friction

Writing
- Reading Assignment: *DK Encyclopedia of Science* pg. 121 Friction, pg. 122 Gravity
- Additional Research Readings
  - Relativity and Gravity: *KSE* pp. 298-299
  - Friction: *KSE* pp. 308-309
  - Gravitation: *UDIS* pp. 18-19

Dates
- 1630's – Galileo does a series of experiments with a marble and a series of differently-shaped tracks, which leads to the discovery of a retarding force called friction.
- 1955 – Christopher Cockerell invents the hovercraft, which uses a cushion of air that allows a vehicle to move without friction.

Schedules for Week 2

Two Days a Week

Day 1	Day 2
☐ Do the "How does friction affect movement?" experiment, and then fill out the experiment sheet on SG pp. 26-27 ☐ Define air resistance, friction, gravity, and terminal velocity on SG pg. 16 ☐ Enter the dates onto the date sheets on SG pp. 8-13	☐ Read pp. 121 and 122 from *DK EOS,* and then discuss what was read ☐ Color and label the "Types of Friction" sketch on SG pg. 25 ☐ Prepare an outline or narrative summary; write it on SG pp. 28-29

Supplies I Need for the Week

- ✓ Force Meter from last week
- ✓ Small wooden block (like a Jenga block), Eye-hook screw
- ✓ Sandpaper, Felt, Foil , Spray oil
- ✓ Tape measure

Things I Need to Prepare

Five Days a Week

Day 1	Day 2	Day 3	Day 4	Day 5
☐ Do the "How does friction affect movement?" experiment, and then fill out the experiment sheet on SG pp. 26-27 ☐ Enter the dates onto the date sheets on SG pp. 8-13	☐ Read pp. 121 and 122 from *DK EOS,* and then discuss what was read ☐ Write an outline on SG pg. 28	☐ Define air resistance, friction, gravity, and terminal velocity on SG pg. 16 ☐ Color and label the "Types of Friction" sketch on SG pg. 25	☐ Read one or all of the additional reading assignments ☐ Write a report on what you learned on SG pg. 29	☐ Complete one of the Want More Activities listed **OR** ☐ Study a scientist from the field of Physics

Supplies I Need for the Week

- ✓ Force Meter from last week
- ✓ Small wooden block (like a Jenga block), Eye-hook screw
- ✓ Sandpaper, Felt, Foil , Spray oil
- ✓ Tape measure

Things I Need to Prepare

Additional Information Week 2

Experiment Information

- **Note** – Make sure your students keep the Jenga block with the eye-screw in it for next week's experiment.
- **Introduction** – (*from the Student Guide*) When an object is in forward motion, several forces are acting on it. There is the driving force, which is propelling the object forward. There is weight (or gravity), which pulls the object downward. There is air resistance, which slows the object down. Finally, there is friction. In today's experiment, you are going to act as the driving force for a block as it moves across a track. Then, you are going to use a variety of materials to test how friction affects the motion of the block.
- **Results** – The students should see that more force was needed to pull the block when it was on the felt and sandpaper. They should see that less force was needed to pull the block when it was on the oil-covered foil.
- **Explanation** – Both the felt and the sandpaper increase the amount of friction that acts on the block as it slides over the track. The oil-coated foil reduces the amount of friction that acts on the block as it slides over the track. Friction is a force that opposes the motion of an object as it passes another. So, when friction increases, the object will slow down. Conversely, when friction is decreased, the object will speed up.
- **Troubleshooting** – The following video show how this experiment should be set up:
  - https://m.youtube.com/watch?v=HP8H3HWBrZE
- **Take if Further** – Have the students explore other ways to reduce the friction the block experiences as it moves up the track. Round toothpicks or marbles would both be good ideas to test.

Discussion Questions

Friction, pg. 121

1. How does the roughness of a surface affect the amount of friction? (*The rougher a surface, the stronger the force of friction.*)
2. Why is friction so important? (*Friction is important because without it we would continuously slide throughout life.*)
3. What does a streamlined design do? (*A streamlined design reduces friction so that the object can move more easily.*)

Gravity, pg. 122

1. What two things affect the force of gravity? (*The distance between the objects and the mass of the objects both affect the force of gravity.*)
2. What is the center of gravity? (*The center of gravity is the point at which the weight of an object appears to be located.*)
3. How does gravity cause the tides in the ocean? (*The gravity of the moon pulls on the ocean on the side of the Earth that is closest, causing it to bulge out. As the two bodies rotate around each other, the strength of the force changes, causing the tides.*)

Want More

- **Galileo's Tracks** – Have the students study the effect of friction using a marble track, just like

Galileo did. Have the students set up a track and send the marble down it several times. Each time, have them record the time it takes to get to the bottom. Then, have the student sprinkle some sand or salt all over the track. Have them send the marbles down several more times, recording the time it takes. (*The students should see that the marbles were much slower the second time, due to the amount of friction that was created by the sand or salt on the track.*)

- **Friction Demonstration** – Do the friction simulation from the PhET website.
  - http://phet.colorado.edu/sims/html/gravity-force-lab/latest/gravity-force-lab_en.html
- **Gravity** – Have the students test gravity using several objects from nature. Begin by taking a walk outside with the students. Have them look for several objects in nature that are round and about the same size. The objects should have different weights, such as a piece of fruit, a rock, and a nut. Once you get home, have the students hold each of the round objects in their hands and drop them at the same time. What happened? (*They should see that both of the objects hit the ground at the very same time. If you can do this safely from a porch or balcony that will give you a bit more height, your results will be even more amazing.*)

Sketch Week 2

This week's sketch assignment is a bit different because the material is not covered in the main text. Instead you will need to read one of the following definitions to your students and help them to figure out which of the four diagrams represents that type of friction.

- **Fluid friction** - The force that opposes the motion of an object through a fluid. *(In physics, water or air are both examples of fluids.)*
- **Static friction** – The friction force that acts on objects that are not in motion, keeping them in balance.
- **Rolling friction** – The force caused by the changing shape of the points of contact as an object rolls across a surface. This type of friction is 10 to 100 times less than static or sliding friction.
- **Sliding friction** – The force that opposes the direction of motion of an object as it slides over a surface.

Repeat this process for each definition until their sketch looks like the one pictured.

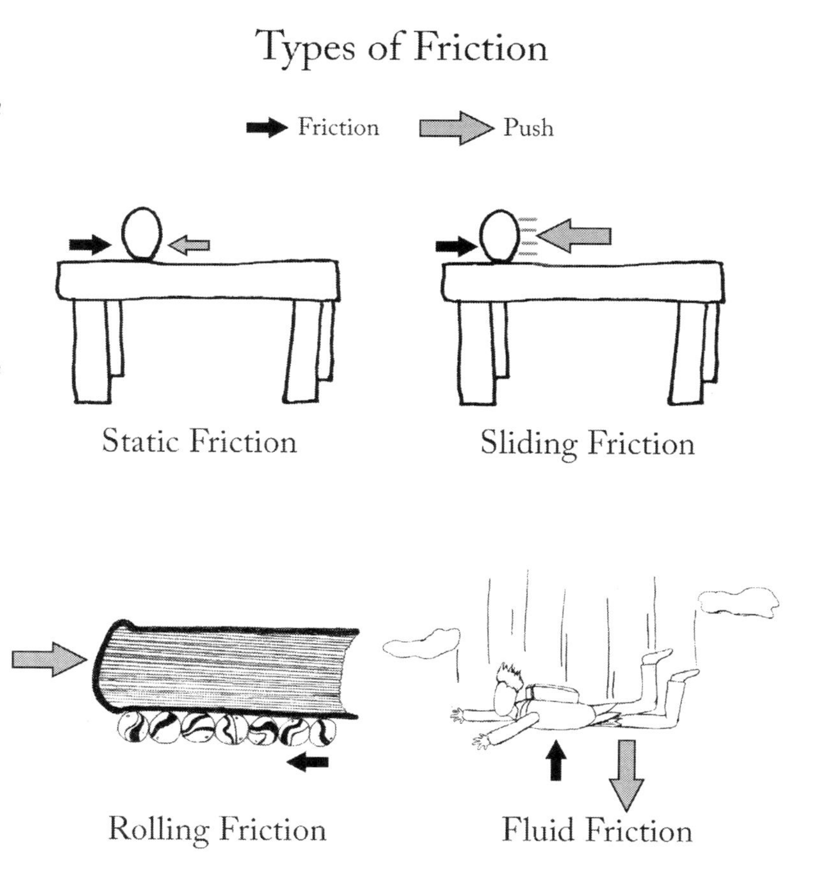

Student Assignment Sheet Week 3
Motion

Experiment: Investigating the Three Laws

Materials

- ✓ Jenga block with the eyehook from last week
- ✓ String
- ✓ 2 Toy cars
- ✓ Egg

Procedure

1. In this experiment, you will be investigating the three laws of motion. Begin by reading the introduction.
2. Test the three laws of motion.
   - **Motion Law # 1** – You will need a Jenga block with the eyehook and string. Tie the string to the block and place it on a smooth surface. Pull the block along with a decent amount of force and then stop suddenly. Observe what happens to the block.
   - **Motion Law # 2** – You will need the two toy cars and a partner for this test. Line up the two cars evenly on a flat surface. Gently push your car forward, while your partner pushes his car forward with a greater force at the same time. Observe what happens to the two cars.
   - **Motion Law # 3** – You will need an egg for this test. Head outside with the egg. Drop the egg onto the pavement from a height of four to five feet and observe what happens.
3. Draw conclusions and complete the experiment sheet.

Vocabulary & Memory Work

- ☐ Vocabulary: inertia, mass, weight, momentum
- ☐ Memory Work — This week, continue working on memorizing Newton's three laws of motion. Also work on memorizing the following equation from Newton's second law:
  - Force (F) = mass (m) • acceleration (a)

Sketch: 3 Laws of Motion

- Label each of the three sketches with the law of motion that they represent. (*See the experiment sheet for a list of the laws.*)

Writing

- Reading Assignment: *DK Encyclopedia of Science* pg. 120 Forces and Motion
- Additional Research Readings
  - Momentum: *KSE* pp. 296-297
  - Dynamics: *UDIS* pp. 12-13

Dates

- 1665 – The plague breaks out in London, which forces Isaac Newton to leave Trinity College in Cambridge. He goes home and spends the next two years working on his book, *Principia*, in which he shares his three laws of motion.

Schedules for Week 3

Two Days a Week

Day 1	Day 2
☐ Do the "Investigating the Three Laws" experiment, and then fill out the experiment sheet on SG pp. 32-33 ☐ Define inertia, mass, weight, and momentum on SG pg. 17 ☐ Enter the dates onto the date sheets on SG pp. 8-13	☐ Read pp. 120 from *DK EOS*, and then discuss what was read ☐ Color and label the "3 Laws of Motion" sketch on SG pg. 31 ☐ Prepare an outline or narrative summary; write it on SG pp. 34-35

Supplies I Need for the Week
- ✓ Jenga block with the eyehook from last week
- ✓ String
- ✓ 2 Toy cars
- ✓ Egg

Things I Need to Prepare

Five Days a Week

Day 1	Day 2	Day 3	Day 4	Day 5
☐ Do the "Investigating the Three Laws" experiment, and then fill out the experiment sheet on SG pp. 32-33 ☐ Enter the dates onto the date sheets on SG pp. 8-13	☐ Read pp. 120 from *DK EOS*, and then discuss what was read ☐ Write an outline on SG pg. 34	☐ Define inertia, mass, weight, and momentum on SG pg. 17 ☐ Color and label the "3 Laws of Motion" sketch on SG pg. 31	☐ Read one or all of the additional reading assignments ☐ Write a report on what you learned on SG pg. 35	☐ Complete one of the Want More Activities listed **OR** ☐ Study a scientist from the field of Physics

Supplies I Need for the Week
- ✓ Jenga block with the eyehook from last week
- ✓ String
- ✓ 2 Toy cars
- ✓ Egg

Things I Need to Prepare

Additional Information Week 3

Notes

- **Inertia vs. Momentum** – Inertia and momentum are often confused or used interchangeably. However, the two are quite different. The amount of inertia force an object experiences is only based on the object's mass, whereas the force of momentum an object feels is dependent upon its mass and its speed. In other words, inertia is how much something resists changes in motion, while momentum increases or decreases with motion.

Experiment Information

- **Note** – This experiment is meant to give your students the chance to see Newton's laws of motion in action. The experiment sheet looks a bit different because of this. So, there is no hypothesis and a separate procedure and observations section for each test.
- **Introduction** – (*from the Student Guide*) Isaac Newton built on Galileo's work on friction and motion through number of experiments. These tests led to his development of the three laws of motion. The laws state:
  1. An object will not move, unless a force like a push or pull moves it. Once it is moving, an object will not stop moving in a straight line unless it's forced to change.
  2. The greater the force on an object, the greater the change in its motion. The greater the mass of an object, the greater the force needed to change its motion.
  3. For every reaction, there is an equal but opposite reaction.

  In today's experiment, you are going to do three tests where you will see each of the laws in action.
- **Results** – For test one, the students should see that when they stopped suddenly, the block continued to move until the string went taut. Once the string was tight, the block to bounced back a bit and eventually stopped. For test two, the students should see that the car that was pushed with greater force moved faster and farther than the other car. For test three, the students should see that the egg cracked and splattered on the pavement.
- **Explanation** – In test one, the students are looking at inertia from the first law of motion. The block began to move because the student pulled the string attached to it. When the student stops suddenly, the block continues to move because the force of the sudden stop in the string has not acted on it yet. However, when then string goes taut, the force is strong enough to change the motion of the block and eventually stop it. In test two, the students should see that the force that acts on the second car is greater than the force that acts on the first one. This causes a greater change in the second car's motion, which means it goes faster and farther. In test three, the egg reaches its terminal velocity before it hits the pavement. It is moving with such a force that the action of the sudden stop causes an equally violent reaction. This reaction has the force to break apart the egg and splatter it on the pavement.

Discussion Questions

1. What did Aristotle believe about motion? (*Aristotle believed that for an object to move it had to be pushed by a force. He also believed that the object would only stop when the force was removed.*)
2. What did Galileo learn about motion? (*Galileo learned that a force was only need to start, stop, and accelerate an object. He found that if the object was already in motion, no force*

was needed to keep it in motion.)

3. What did Isaac Newton discover about motion? (*Isaac Newton discovered three laws about motion. The first says that an object will stay still or keep on the same path at a constant speed unless a force pushes or pulls it. The second law says that the greater the force that acts on an object, the greater the change in movement. The third law says that for every action, there is an equal and opposite reaction.*)

Want More

- **Egg Drop Carrier** – Have the students test how to slow down the inertia and momentum of a falling egg so that the "reaction" from the third law of motion doesn't result in a cracked egg. You will need a raw egg, various shock absorbing materials (such as cotton balls, newspaper, packing peanuts or fabric), a 1 quart plastic container (the type that fruit is typically packed in), and masking tape. Have the student begin by examining the different shock absorbing materials you have and using it to fill the 1 quart container in such a way that the egg will be protected as it falls. Be sure to have the egg on hand so that you can measure it and make sure the egg fits in the remaining space. Have the students tape across the top of the container to hold in the egg and the materials. Now, have them hold their containers over their heads and drop it. Observe what happens. Did the egg crack? (*As the egg drops, its speed increases, which causes the force of momentum to increase. When it hits the ground, an equal shock force (Newton's 3rd Law of Motion) is sent back into the egg. If the egg was not protected, it would surely crack, like it did in the experiment from earlier this week. However, in this case, the materials surrounding the egg absorb most of the shock force, so that the force that the egg eventually felt was minimal.*)
- **Second Law Worksheet** – Have the students practice calculating acceleration, mass, and force using the second law of motion using the worksheet in the Appendix on pg. 251. Answers: 1. $a = 8\ m/s^2$ 2. $F = 160{,}000\ N$ 3. $m = 25\ kg$

Sketch Week 3

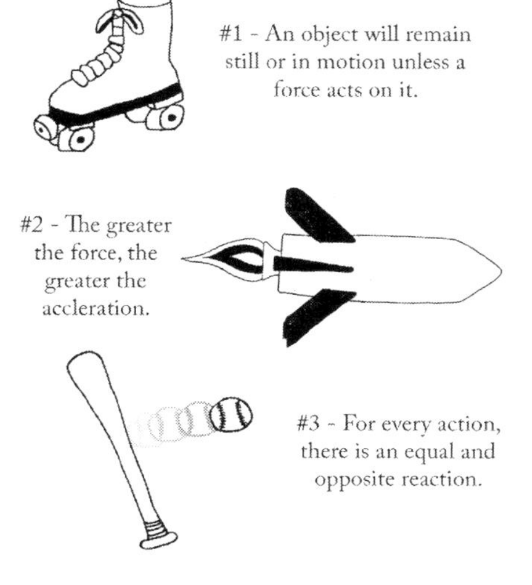

Student Assignment Sheet Week 4
Speed and Acceleration

Experiment: Will the height of the ramp affect a car's speed?

Materials

- ✓ Cardboard or plastic track
- ✓ Blocks or thick books
- ✓ Toy car
- ✓ Stopwatch

Procedure

1. Read the introduction to the experiment and answer the question in the hypothesis section.
2. Use the cardboard (or plastic track) to build a track that is 1 meter long. Use the blocks (or books) to prop the track up so that one end of it is 15 centimeters higher than the other. Now, use the stopwatch to measure the time it takes for the car to go from the top of the track to the end. Record the time on the experiment sheet and then repeat two more times for a total of three trials at 15 centimeters.
3. Next, add several more blocks (or books) to prop the track up so that one end of it is 30 centimeters higher than the other. Use the stopwatch to measure the time it takes for the car to go from the top of the track to the end. Record the time on the experiment sheet and then repeat two more times for a total of three trials at 30 centimeters.
4. Finally, add few more blocks (or books) to prop the track up so that one end of it is 45 centimeters higher than the other. Use the stopwatch to measure the time it takes for the car to go from the top of the track to the end. Record the time on the experiment sheet and then repeat two more times for a total of three trials at 45 centimeters.
5. Draw conclusions and complete the experiment sheet.

Vocabulary & Memory Work

- ☐ Vocabulary: acceleration, speed, velocity
- ☐ Memory Work—This week, continue working on memorizing Newton's three laws of motion. Also work on memorizing the following equations for velocity and acceleration:
  - Speed (v) = total distance (d) / total time (t)
  - Acceleration (a) = change in velocity ($v_f - v_i$) /total time (t)

Sketch: Acceleration Graph

- Label the following parts of the line on the graph – constant acceleration, constant speed, and constant deceleration

Writing

- Reading Assignment: *DK Encyclopedia of Science* pg. 118 Speed, pg. 119 Acceleration
- Additional Research Readings
  - Motion: *UDIS* pp. 10-11

Dates

- 1905 – Albert Einstein publishes his theory of relativity, which is the basis for many of the ideas we have about our universe.

Schedules for Week 4

Two Days a Week

Day 1	Day 2
☐ Do the "Will the height of the ramp affect a car's speed?" experiment, and then fill out the experiment sheet on SG pp. 38-39 ☐ Define acceleration, speed, and velocity on SG pg. 17 ☐ Enter the dates onto the date sheets on SG pp. 8-13	☐ Read pp. 118-119 from *DK EOS,* and then discuss what was read ☐ Color and label the "Acceleration Graph" sketch on SG pg. 37 ☐ Prepare an outline or narrative summary; write it on SG pp. 40-41

Supplies I Need for the Week

- ✓ Cardboard or plastic track
- ✓ Blocks or thick books
- ✓ Toy car
- ✓ Stopwatch

Things I Need to Prepare

Five Days a Week

Day 1	Day 2	Day 3	Day 4	Day 5
☐ Do the "Will the height of the ramp affect a car's speed?" experiment, and then fill out the experiment sheet on SG pp. 38-39 ☐ Enter the dates onto the date sheets on SG pp. 8-13	☐ Read pp. 118-119 from *DK EOS,* and then discuss what was read ☐ Write an outline on SG pg. 40	☐ Define acceleration, speed, and velocity on SG pg. 17 ☐ Color and label the "Acceleration Graph" sketch on SG pg. 37	☐ Read one or all of the additional reading assignments ☐ Write a report on what you learned on SG pg. 41	☐ Complete one of the Want More Activities listed **OR** ☐ Study a scientist from the field of Physics

Supplies I Need for the Week

- ✓ Cardboard or plastic track
- ✓ Blocks or thick books
- ✓ Toy car
- ✓ Stopwatch

Things I Need to Prepare

Additional Information Week 4

Notes

- **Acceleration** – Often acceleration refers to an object that is speeding up, which is true to a point. In scientific terms, acceleration is related to velocity, not speed. Remember that velocity is a vector quantity, meaning it has both speed and direction. So, in physics, acceleration can be speeding up, slowing down, or just a change in direction.
- **Scalar quantity vs. vector quantity** – A scalar quantity is a quantity with only one magnitude. Mass, time, and inertia are all examples of scalar quantities. A vector quantity is a quantity with both magnitude and direction. Force, momentum, and velocity are all examples of vector quantities.

Experiment Information

- **Introduction –** (*from the Student Guide*) We know from Newton's laws of motion that a force must act on a vehicle to get it moving. We also know that the greater the force, the quicker the vehicle will go. In today's experiment, you are going to measure the time it takes for a toy car to go down a given length of track. You will also vary the height of the track to see how the different angles increase or decrease force of gravity affecting the speed of the car.
- **Results** – The students should see that higher the track, the faster the average speed of the car.
- **Explanation –** For the car to move down the track a force must either push or pull. In this experiment, the force of gravity overcomes the forces of friction and inertia to send the car heading down the track. As the height of the track increases, the force of gravity that pulls on the car also increases. This causes the speed of the toy car to increase as the height of the track increases.
- **Troubleshooting –** Make sure that the track is smooth, even, and free from debris that might slow down the progress of the car.
- **Take it Further –** Have the students cover the track with sandpaper to increase the friction. Then, have them repeat the experiment to see how the increased friction affects their results.

Discussion Questions

Speed, pg. 118

1. What is the average speed? (*Average speed is the distance traveled divided by the time taken.*)
2. What is instantaneous speed? (*Instantaneous speed is the speed at which an object is traveling at a given instant.*)
3. How are speed and velocity related? (*Velocity is the measurement of speed in a particular direction.*)

Acceleration, pg. 119

1. What is acceleration? (*Acceleration is a measure of how quickly velocity increases.*)
2. What is deceleration? (*Deceleration is a negative acceleration or a measure of how quickly velocity decreases.*)

Want More

- **Acceleration worksheet –** Have the students practice calculating acceleration using the worksheet in the Appendix on pg. 252.
  Answers
  1. 1.43 m/s^2
  2. 5 m/s^2
  3. 5.2 m/s^2

Sketch Week 4

Acceleration Graph

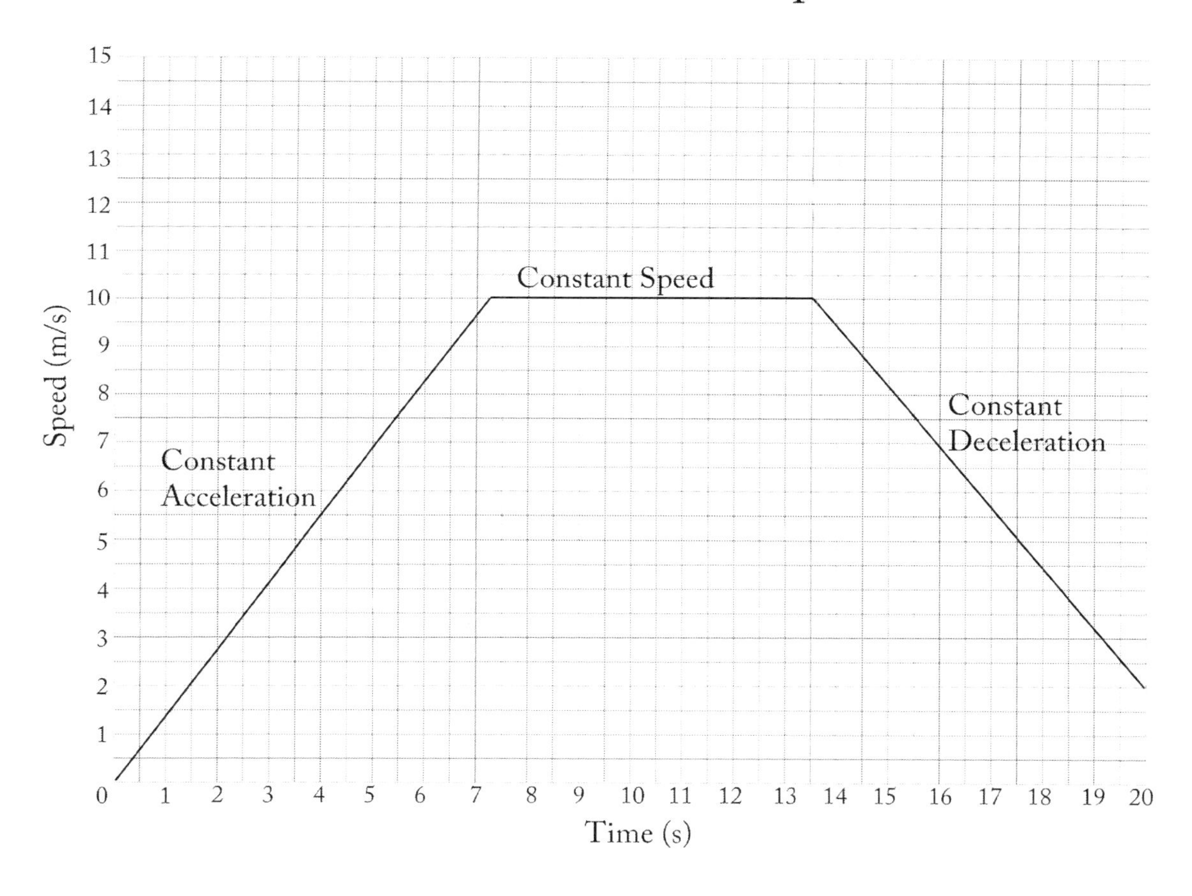

Unit 1 Forces and Motion
Unit Test Answers

Vocabulary Matching

1. B
2. E
3. K
4. A
5. F
6. M
7. H
8. C
9. I
10. L
11. N
12. D
13. J
14. O
15. G

True or False

1. True
2. False (*An object is balanced when the forces acting on it cancel each other out, which produces a zero resultant.*)
3. False (*A streamlined design reduces friction so that the object can move more easily.*)
4. True
5. True
6. False (*Isaac Newton discovered the three laws about motion.*)
7. False (*Average speed is the distance traveled divided by the time taken.*)
8. True

Short Answer

1. The resultant force is the overall result of two or more forces acting on an object.
2. The distance between the objects and the mass of the objects both affect the force of gravity.
3. The first says: an object will stay still or keep on the same path at a constant speed unless a force pushes or pulls it. The second law says: the greater the force that acts on an object, the greater the change in movement. The third law says: for every action there is an equal and opposite reaction.
4. Velocity is the measurement of speed in a particular direction.
5. Force (F) = mass (m) • acceleration (a)

Unit 1 Forces and Motion
Unit Test

Vocabulary Matching

1. Balance ____
2. Force ____
3. Force field ____
4. Newton ____
5. Air resistance ____
6. Friction ____
7. Gravity ____
8. Terminal velocity ____
9. Inertia ____
10. Mass ____
11. Momentum ____
12. Weight ____
13. Acceleration ____
14. Speed ____
15. Velocity ____

A. The measurement of force; 1 Newton (N) is the force is takes to move a one kilogram object at 1 meter per second squared (1 N = 1 kg • 1 m/s^2).

B. A state of equilibrium when the forces acting on an object cancel each other out; also known as a zero resultant force.

C. The point at which the force acting on an object of air resistance is equal to the force of gravity acting on the object.

D. The force with which an object's mass is pulled toward the center of the Earth.

E. A push or pull that acts on an object.

F. The force that air exerts on an object as it falls.

G. The speed of an object in a particular direction.

H. The force that acts between two masses; it is an attractive force.

I. The tendency of an object to resist a change in its motion.

J. A change in an objects speed, direction, or both.

K. The area in which a force can be felt.

L. The amount of matter in an object.

M. A force that opposes the motion of objects that touch as they move past each other.

N. The tendency of an object to keep moving until a force stops it.

O. The ratio of the distance an object moves to the amount of time the object moves.

True or False

1. ____________________ Contact forces are only produced when one object touches another. Non-contact forces can pull objects without touching them.

2. ______________________ An object is balanced when the forces acting on it add to each other which produces a positive resultant force.

3. ______________________ A streamlined design increases friction so that the object can move more easily.

4. ______________________ The center of gravity is the point at which the weight of an object appears to be located.

5. ______________________ Aristotle believed that the object would only stop when the force was removed.

6. ______________________ Galileo discovered a great deal about motion, including the three laws of motion.

7. ______________________ Average speed is the speed at which an object is traveling at a given instant.

8. ______________________ Acceleration is a measure of how quickly velocity increases or decreases.

Short Answer

1. What is a resultant force?

2. What two things affect the force of gravity between two objects?

3. What are Newton's three laws of motion?

4. How are speed and velocity related?

5. Write the equation that relates force, mass, and acceleration that comes from Newton's second law of motion.

Physics Unit 2

Energy

Unit 2 Energy
Overview of Study

Sequence of Study

Week 5: Energy
Week 6: Pressure
Week 7: Energy Sources
Week 8: Simple Machines

Materials by Week

Week	Materials
5	Goldfish cracker, Small marshmallow, Piece of lettuce, Piece of bacon fat, Aluminum pan, Matches, Safety glasses, Bucket of water
6	2-Liter Soda bottle, 2 Cans – one large, one small, Screw, Water, Piece of clay, Cup measure, Tape measure
7	Foil, Black construction paper, Small cardboard box, Plastic wrap, Tape, Marshmallow, Small glass dish (one that will fit inside the box)
8	Materials will vary depending upon the simple machine the student chooses to build

Vocabulary for the Unit

1. **Energy** – The ability to do work.
2. **Kinetic energy** – The energy of an object in motion; it depends upon the object's mass and speed.
3. **Potential energy** – The energy that an object has stored; it depends upon the object's weight and height.
4. **Work** – The transfer of energy that occurs when a force moves or changes an object.
5. **Fluid** – A substance that assumes the shape of the container it is in, such as a liquid or gas.
6. **Pressure** – The amount of force pushing on a giving area.
7. **Non-renewable energy resources** – Energy sources that exist in limited quantities, such as oil and coal.
8. **Renewable energy resources** – Energy sources that can be replaced in a relatively short period of time, such as wind and solar.
9. **Fulcrum** – The point on which a lever rests or is supported and on which it pivots.
10. **Input force** – The force you put into a machine.
11. **Output force** – The force that a machine exerts on an object.

Memory Work for the Unit

Types of Energy

1. **Mechanical energy** – The energy associated with the motion and position of an object. It is the sum of an object's kinetic and potential energy.
2. **Chemical energy** – The energy that is stored in chemical bonds. It is released when these bonds are broken in a chemical reaction.
3. **Thermal energy** – The energy that flows from one place to another due to changes in temperature. It is also known as heat.
4. **Electrical energy** – The energy associated with electrical charges.
5. **Electromagnetic energy** – The energy that travels through space in the form of waves, such as x-rays, light, and sound.
6. **Nuclear energy** – The energy stored in an atomic nucleus.

Equations

- Work Equation

  $W = F \bullet d$

  "W" stands for work.
  "F" stands for force.
  "d" stands for distance.

- Pressure Equation

  $P = \frac{F}{a}$

  "P" stands for pressure.
  "F" stands for force.
  "a" stands for area.

Notes

Student Assignment Sheet Week 5
Energy

Experiment: Do different types of food contain different amounts of energy?

Materials

- ✓ Goldfish cracker
- ✓ Small marshmallow
- ✓ Piece of lettuce
- ✓ Piece of bacon fat
- ✓ Aluminum pan
- ✓ Matches
- ✓ Safety glasses
- ✓ Bucket of water

Procedure

****Caution—Be sure to do this experiment in a well-ventilated area with a fire extinguisher close at hand.****

1. Read the introduction to the experiment and answer the question for the hypothesis section.
2. Head outside and set the aluminum pan on a concrete or tile surface or inside a grill. Place the cracker in the aluminum pan and have an adult use the match to light the cracker. Time how long it takes for the cracker to burn and write this on your experiment sheet.
3. Use the bucket of water to extinguish any smoldering embers. Then, clean and dry the aluminum pan.
4. Now, repeat the procedure from step 2 and 3 for the marshmallow, lettuce, and bacon fat. (**Note**—*Make sure that your samples are similar in size, so that your results are more accurate*.)
5. Draw conclusions and complete the experiment sheet.

Vocabulary & Memory Work

- ❒ Vocabulary: energy, potential energy, kinetic energy, work
- ❒ Memory WorkBegin working on memorizing the types of energy (see Unit Overview Sheet) along with the following equation:
  - Work (W) = Force (F) • Distance (d)

Sketch: Energy Diagram

- Label the following – Energy, kinetic energy, energy from motion; potential energy, energy from gravity, energy from a stretched or compressed object, energy from an electrical charge, energy from the pull of a magnet

Writing

- Reading Assignment: *DK Encyclopedia of Science* pp. 132-133 Work and Energy
- Additional Research Readings
  - Potential and Kinetic Energy: *KSE* pp. 292-293
  - Energy: *UIDS* pp. 8-9

Dates

- 1818-1889 – James Joule lives. He is credited with discovering that work produces heat, which is a form of energy.

Schedules for Week 5

Two Days a Week

Day 1	Day 2
☐ Do the "Do different types of food contain different amounts of energy?" experiment, and then fill out the experiment sheet on SG pp. 48-49 ☐ Define energy, potential energy, and kinetic energy, work on SG pg. 44 ☐ Enter the dates onto the date sheets on SG pp. 8-13	☐ Read pp. 132-133 from *DK EOS,* and then discuss what was read ☐ Color and label the "Energy Diagram" sketch on SG pg. 47 ☐ Prepare an outline or narrative summary, write it on SG pp. 50-51
Supplies I Need for the Week ✓ Goldfish cracker, Small marshmallow, Piece of lettuce, Piece of bacon fat ✓ Aluminum pan, Matches ✓ Safety glasses, Bucket of water	
Things I Need to Prepare	

Five Days a Week

Day 1	Day 2	Day 3	Day 4	Day 5
☐ Do the "Do different types of food contain different amounts of energy?" experiment, and then fill out the experiment sheet on SG pp. 48-49 ☐ Enter the dates onto the date sheets on SG pp. 8-13	☐ Read pp. 132-133 from *DK EOS,* and then discuss what was read ☐ Write an outline on SG pg. 50	☐ Define energy, potential energy, and kinetic energy, work on SG pg. 44 ☐ Color and label the "Energy Diagram" sketch on SG pg. 47	☐ Read one or all of the additional reading assignments ☐ Write a report on what you learned on SG pg. 51	☐ Complete one of the Want More Activities listed **OR** ☐ Study a scientist from the field of Physics
Supplies I Need for the Week ✓ Goldfish cracker, Small marshmallow, Piece of lettuce, Piece of bacon fat ✓ Aluminum pan, Matches ✓ Safety glasses, Bucket of water				
Things I Need to Prepare				

Additional Information Week 5

Note

- **Gravitational vs. Elastic Potential Energy** – There are two types of potential energy. The first is gravitational potential energy which is dependent on the object's height. In other words, the higher an object is, the more potential energy it has due to the effect of gravity. The second is elastic potential energy, which is dependent on whether an object is stretched or compressed. In other words, this type of potential energy increases the more an object is stretched or compressed.
- **Types of Energy** – Most types of energy are either kinetic, potential, or a combination of the two. All other forms of energy are in fields, such as electromagnetic waves.

Experiment Information

- **Introduction** – (*from the Student Guide*) As we break down the food we eat, it releases energy in the form of calories. Some of the chemicals in food are broken down quickly, releasing a burst of energy. Other chemicals take a bit more work to break down, so they release energy slowly over time. These reactions work together to fuel our bodies' processes over time. In today's experiment, you will be testing different types of food to see how much energy they release.
- **Results –** The student should see that the marshmallow burned the quickest, followed by the lettuce, the goldfish cracker, and finally the bacon.
- **Explanation –** The marshmallow has a quick sugar-rush of energy, which burns and then quickly dies out. The lettuce has fiber and a bit of sugar, which burns slower than the marshmallow. However, since lettuce is mostly water, it doesn't burn all that long. The goldfish cracker has a mixture of sugar, starch, and fiber, which keeps it burning longer than the two previous samples. The bacon fat is full of lipids and proteins that burn for a long time. Just like the food we eat, vegetables give us the fuel our bodies needs, but we have to eat a lot of bulk to survive on only vegetables. Sweets, like marshmallows, give our bodies a quick punch of energy that doesn't last very long. Carbohydrates rich in fiber fuel our bodies for a much longer period of time than sugar. Fats and proteins provide our bodies with a long-lasting source of energy. Since our bodies need more than just fuel, it is best to have a balance of fruits and vegetables, carbohydrates, along with fats and proteins.
- **Take it Further –** Have the students compare even more types of foods to see if there is a difference.

Discussion Questions

1. What happens when work is done? (*When work is done, energy is changed from one form to another*.)
2. How are work and energy related? (*Energy is the ability to do work and work cannot be done without energy*.)
3. What are some common forms of energy? (*Several common forms of energy include chemical, heat, light, sound, nuclear, and electrical*.)
4. What are the two units of measurement for energy? (*Energy can be measured in Joules or calories*.)

5. What is the difference between kinetic and potential energy? (*Kinetic energy is the energy of motion, but potential energy is the energy that an object has because of its position.*)

Want More

- **Work Worksheet** – Have your students practice calculating work using the worksheet in the Appendix on pg. 253.
  Answers
  1. The man has done 200 J of work.
  2. The person has done 240 J of work.
  3. The hot air has done 630,000 J of work.
  4. The winch has done 1050 J of work.

Sketch Week 5

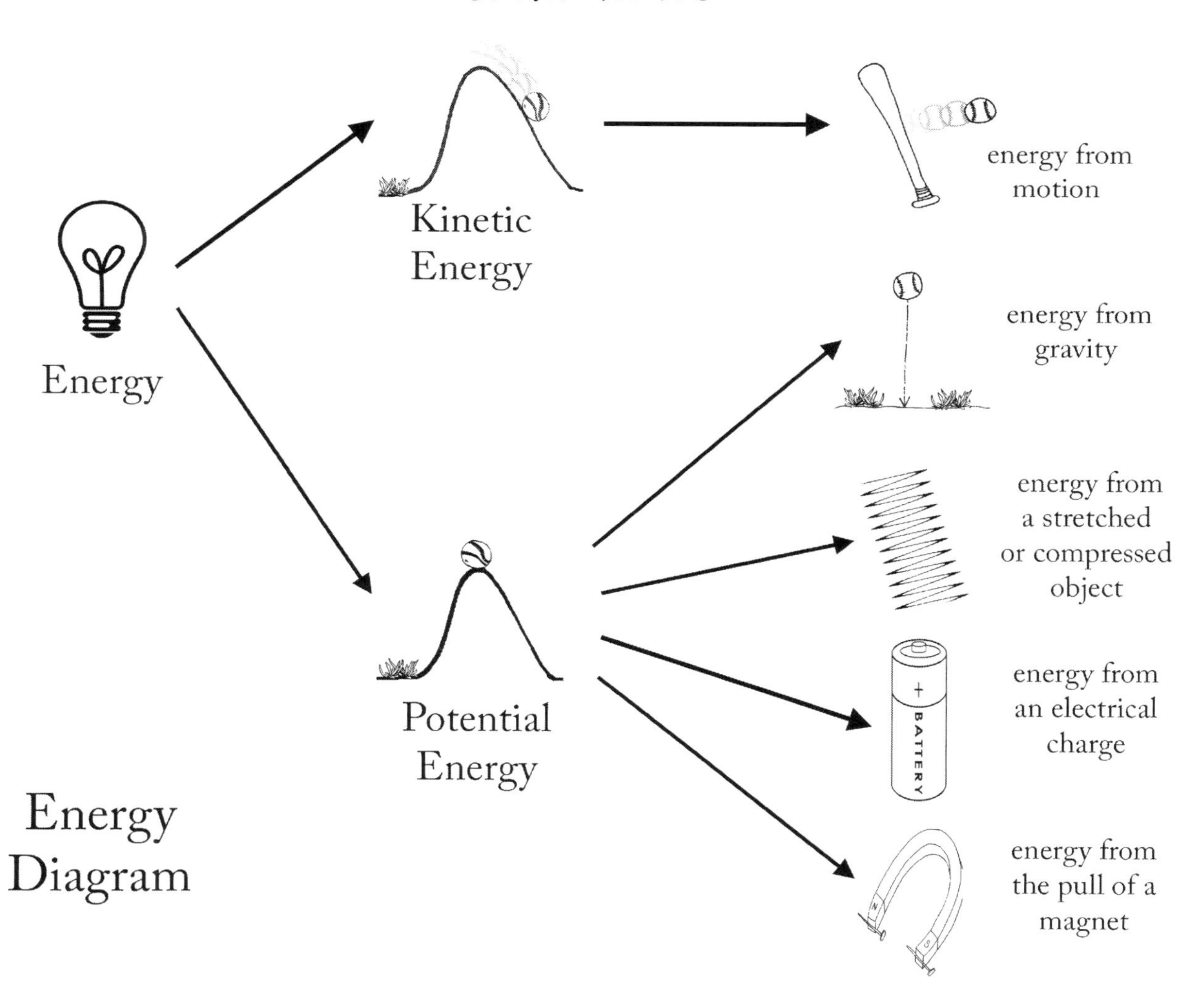

Student Assignment Sheet Week 6
Pressure

Experiment: What affects the pressure exerted by a fluid?

Materials

- ✓ 2-Liter Soda bottle
- ✓ 2 Cans – one large, one small
- ✓ Screw
- ✓ Water
- ✓ Piece of clay
- ✓ Cup measure
- ✓ Tape measure

Procedure

1. Read the introduction to the experiment and answer the question for the hypothesis section.
2. Perform the two pressure tests.
    - **Pressure Test #1** – Fill the small tin can with water and then use a cup measure to determine how much water the can holds. Use the screw to poke a hole in the bottom of the small tin can about ½ inch (1.25 cm) from the bottom. Put a piece of clay over the hole to prevent any water spilling out before filling the can with water. Lay out the tape measure on a surface that you don't mind getting wet, preferably outdoors. Place the can at the first mark on the tape measure. Quickly remove the clay plug. Measure how far the water goes and record it on the experiment sheet. Repeat this process two more times and then average the distances. Repeat steps 2 and 3 using the larger can.
    - **Pressure Test #2** – Use the screw to poke three holes in the 2-Liter bottle – one about 1 inch (2.5 cm) from the bottom, one at the middle of the bottle, and one 2 inches (5 cm) from the top of the bottle. Plug each of the holes with clay and the fill the bottle to the very top. Place the full 2-Liter bottle at the first mark on the tape measure. Quickly remove the clay plugs. Measure how far the water goes from each of the holes and record it on the experiment sheet. Repeat this process two more times and then average the distances.
3. Draw conclusions and complete the experiment sheet.

Vocabulary & Memory Work

- ☐ Vocabulary: fluid, pressure
- ☐ Memory Work—Continue to work on memorizing the different types of energy along with the following equation:
    - Pressure (P) = $\frac{\text{Force (F)}}{\text{Area (a)}}$

Sketch: Pressure Relationships

- Label the following – Pressure and force relationship, increase force, increase pressure, decrease force, decrease pressure; pressure and area relationship, decrease area, increase pressure, increase area, decrease pressure
- Fill in the equation at the bottom of the sketch.

Writing

- Reading Assignment: *DK Encyclopedia of Science* pg. 127 Pressure
- Additional Research Readings
    - Pressure: *KSE* pg. 311, *UIDS* pg. 25; Fluids: *KSE* pg. 310

Dates

- 1643 – Evangelista Torricelli invents the mercury barometer.

Schedules for Week 6

Two Days a Week

Day 1	Day 2
☐ Do the "What affects the pressure exerted by a fluid?" experiment, and then fill out the experiment sheet on SG pp. 54-55 ☐ Define fluid and pressure on SG pg. 44 ☐ Enter the dates onto the date sheets on SG pp. 8-13	☐ Read pg. 127 from *DK EOS,* and then discuss what was read ☐ Color and label the "Pressure Relationships" sketch on SG pg. 53 ☐ Prepare an outline or narrative summary; write it on SG pp. 56-57

Supplies I Need for the Week

✓ 2-Liter Soda bottle, 2 Cans – one large, one small
✓ Screw, Water, Piece of clay
✓ Cup measure, Tape measure

Things I Need to Prepare

Five Days a Week

Day 1	Day 2	Day 3	Day 4	Day 5
☐ Do the "What affects the pressure exerted by a fluid?" experiment, and then fill out the experiment sheet on SG pp. 54-55 ☐ Enter the dates onto the date sheets on SG pp. 8-13	☐ Read pg. 127 from *DK EOS,* and then discuss what was read ☐ Write an outline on SG pg. 56	☐ Define fluid and pressure on SG pg. 44 ☐ Color and label the "Pressure Relationships" sketch on SG pg. 53	☐ Read one or all of the additional reading assignments ☐ Write a report on what you learned on SG pg. 57	☐ Complete one of the Want More Activities listed **OR** ☐ Study a scientist from the field of Physics

Supplies I Need for the Week

✓ 2-Liter Soda bottle, 2 Cans – one large, one small
✓ Screw, Water, Piece of clay
✓ Cup measure, Tape measure

Things I Need to Prepare

Additional Information Week 6

Experiment Information

- **Introduction –** (*from the Student Guide*) A fluid, such as a liquid or a gas, fills in the available space within a container. The fluid exerts pressure on the surfaces of the container throughout the space it occupies. The amount of that pressure depends upon several factors. In today's experiment, you will be looking at two factors – area and force – to see how they affect the pressure a fluid exerts in a container.
- **Results –** In the first pressure test, the students should see that the water in the smaller can went further than the water in the large can. In the second test, the students should see that the water coming out of the bottom hole went the farthest, followed by the middle hole, and then the top hole.
- **Explanation –** In the first test, the students are examining the effect of area on pressure. The force exerted by the water remains the same because they used the same amount of water for each test. Area and pressure are inversely proportional – the greater the area, the lower the pressure. This is why the water that shot out of the smaller can with the lesser area went further. In the second test, the students are testing the effect of force on pressure. Within the 2-Liter bottle, the force exerted by the water increases with depth. So, the force the water exerts is greatest at the bottom of the bottle. Since force and pressure are directly related – the greater the force, the greater the pressure. For this reason, the water out of the bottom hole in the bottle travels the farthest.
- **Troubleshooting –** Here are tips for each of the pressure tests from the experiment:
  - **Pressure Test #1** – The cans the students use in this test need to have different radiuses. The small can should have a smaller radius than the larger one. In other words, the students should choose a tomato paste can and a soup can, not a tuna can and soup can.
  - **Pressure Test #2** – The students may want to repeat this test several times so that they can measure just one of the distances at a time.
- **Take it Further –** Have the students repeat the experiment, only this time have them use different sizes of screws to see if the size of the hole makes a difference in their results. (*The students should see that the smaller the hole, the farther the water goes. This is because the force stays the same, but the area decreases, which causes the pressure to increase.*)

Discussion Questions

1. How are pressure, force, and area related? (*If you spread force over a large area, you will reduce the amount of pressure exerted. If you concentrate the force over a small area, you will increase the amount of pressure exerted.*)
2. What happens to air pressure as you go high up? (*Air pressure decreases as the altitude increases.*) Why? (*This happens because as you increase your altitude, there is less and less air pressing on you.*)
3. What happens to water pressure as you dive deep under water? (*Water pressure increases the deeper you go under water.*) Why? (*This happens because as you increase your depth under water, there is more and more water pressing down on you.*)
4. Why do we not feel the pressure of the surrounding air on our bodies? (*We do not feel the pressure of the surrounding air on our bodies because the fluids inside our bodies exert as*

much pressure as the surrounding air.)

Want More

- **Forces in Fluids** – Have the students read *DK Encyclopedia of Science* pg. 128 (Forces in Fluid). Have them share with you what Pascal's and Bernoulli's principles state. (*Pascal's principle states that fluids transmit pressure equally in all directions. Bernoulli's principle states that a fast-moving fluid has a lower pressure than a slow-moving one.*)
- **Pressure Worksheet** – Have your students practice calculating pressure using the worksheet in the Appendix on pg. 254.
  Answers
  1. 2500 Pa
  2. 62.5 Pa
  3. 1600 Pa

Sketch Week 6

Pressure Relationships

Pressure and Force relationship

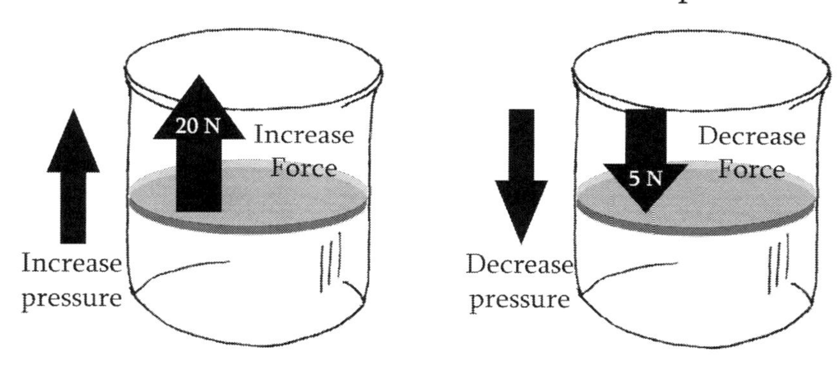

Pressure and Area relationship

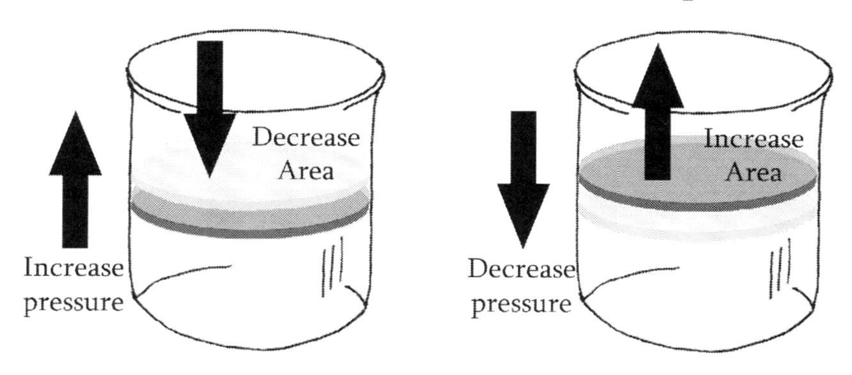

$$\text{Pressure} = \frac{\text{Force}}{\text{Area}}$$

Student Assignment Sheet Week 7
Energy Sources

Experiment: Build a Solar Oven

Materials

- ✓ Foil
- ✓ Black construction paper
- ✓ Small cardboard box (about 9" by 12")
- ✓ Plastic wrap
- ✓ Tape
- ✓ Marshmallow
- ✓ Small glass dish (one that will fit inside the box)

Procedure

1. Read the introduction to the experiment.
2. Place the black construction paper on the bottom of your box. This will serve to absorb the sunlight.
3. Tape the flaps down. Then, cover the sides with the foil so that the shiny side is facing out. The foil will help to reflect the sunlight and focus it on cooking the marshmallow.
4. Put the glass dish in the center of the bottom of the box and set the marshmallow on it. Cover the top box with plastic wrap; this will help to trap the heat.
5. Now, head outside and set your solar oven in direct sunlight. Observe what happens.
6. Draw conclusions and complete the experiment sheet.

Vocabulary & Memory Work

- ☐ Vocabulary: non-renewable energy resources, renewable energy resources
- ☐ Memory Work—Continue to work on memorizing the different types of energy.

Sketch: Energy Sources Chart

- Create a chart depicting renewable and non-renewable energy sources.

Writing

- Reading Assignment: *DK Encyclopedia of Science* pp. 134-135 Energy Sources
- Additional Research Readings
  - Harnessing Wave and Wind Power: *KSE* pp. 328-331

Dates

- c. 100 – The Romans began to burn coal as a source of energy.
- c. 650 – The Persians began to use windmills as a source of energy.
- 1891 – Hydroelectric power is first demonstrated in Germany.
- 1960 – Turkmenistan builds the first solar thermal power plant.

Schedules for Week 7

Two Days a Week

Day 1	Day 2
☐ Do the "Build a Solar Oven" experiment, and then fill out the experiment sheet on SG pp. 60-61 ☐ Define non-renewable energy resources and renewable energy resources on SG pg. 44 ☐ Enter the dates onto the date sheets on SG pp. 8-13	☐ Read pp. 134-135 from *DK EOS,* and then discuss what was read ☐ Color and label the "Energy Sources Chart" sketch on SG pg. 59 ☐ Prepare an outline or narrative summary; write it on SG pp. 62-63

Supplies I Need for the Week

- ✓ Foil, Black construction paper, Small cardboard box (about 9" by 12")
- ✓ Plastic wrap, Tape, Marshmallow
- ✓ Small glass dish (one that will fit inside the box)

Things I Need to Prepare

Five Days a Week

Day 1	Day 2	Day 3	Day 4	Day 5
☐ Do the "Build a Solar Oven" experiment, and then fill out the experiment sheet on SG pp. 60-61 ☐ Enter the dates onto the date sheets on SG pp. 8-13	☐ Read pp. 134-135 from *DK EOS,* and then discuss what was read ☐ Write an outline on SG pg. 62	☐ Define non-renewable energy resources and renewable energy resources on SG pg. 44 ☐ Color and label the "Energy Sources Chart" sketch on SG pg. 59	☐ Read one or all of the additional reading assignments ☐ Write a report on what you learned on SG pg. 63	☐ Complete one of the Want More Activities listed **OR** ☐ Study a scientist from the field of Physics

Supplies I Need for the Week

- ✓ Foil, Black construction paper, Small cardboard box (about 9" by 12")
- ✓ Plastic wrap, Tape, Marshmallow
- ✓ Small glass dish (one that will fit inside the box)

Things I Need to Prepare

Additional Information Week 7

Experiment Information

- ☞ **Introduction** – (*from the Student Guide*) On Earth, we have many different sources of energy. Some of these sources are not renewable, such as fossil fuels like oil and coal. Some of the sources are renewable, like wind, waves, and the sun. In this experiment, you are going to build a device that will allow you to harness the power of the sun to cook a marshmallow.
- ☞ **Results and Explanation** – The purpose of this experiment was to give the students a chance to see how the sun can be used as an energy source. If they were able to cook their marshmallow, they have accomplished this purpose.
- ☞ **Troubleshooting** – Be sure that the solar oven is in direct sunlight at the peak of the day (between 11am and 3pm).
- ☞ **Take it Further** – Have the students work to improve on the design of the solar oven. (**Ideas**—*The students can make it larger, add flaps to focus the sun, or add different types of insulation to trap more of the heat in the box.*)

Discussion Questions

1. What is the main source of renewable energy on the earth? (*The sun is the main source of renewable energy on the earth.*)
2. What is the difference between renewable and non-renewable energy sources? (*Renewable energy sources will not run out, while non-renewable energy sources will eventually run out.*)
3. How do solar panels work? (*Solar panels use photovoltaic cells to convert the sun's energy into electricity.*)
4. How can wind and water create power? (*Wind and water can be used to turn turbines, which drive generators to create electricity.*)
5. How does a power plant work? (*In a power plant, a furnace burns either coal or oil, which heats up water to form steam. The steam is then used to drive a turbine that is attached to a generator. The generator creates electricity which can be sent to homes and buildings in the surrounding area.*)
6. What are fossil fuels? (*Fossil fuels, like coal, natural gas, and oil, come from the remains of long-dead plants and animals.*)

Want More

- **Research Paper** – Have your students research and learn more about renewable energy sources, such as wind, solar, geothermal, and hydro (water). Then, have the students choose one of the sources as a topic for a one to three page research paper. Their paper can include an explanation of the renewable energy source, how it is currently being used, any problems with using the energy source, and how it compares to non-renewable energy sources.

Sketch Week 7

The students can simply write out the different energy fuel sources, or they can use pictures to depict them. Here are some of the sources they could include:

- **Non-renewable energy sources** – oil, natural gas, coal, uranium
- **Renewable energy sources** – biomass (such as wood or animal droppings), wind, water, solar, geothermal

Here is a sample of what the sketch could look like:

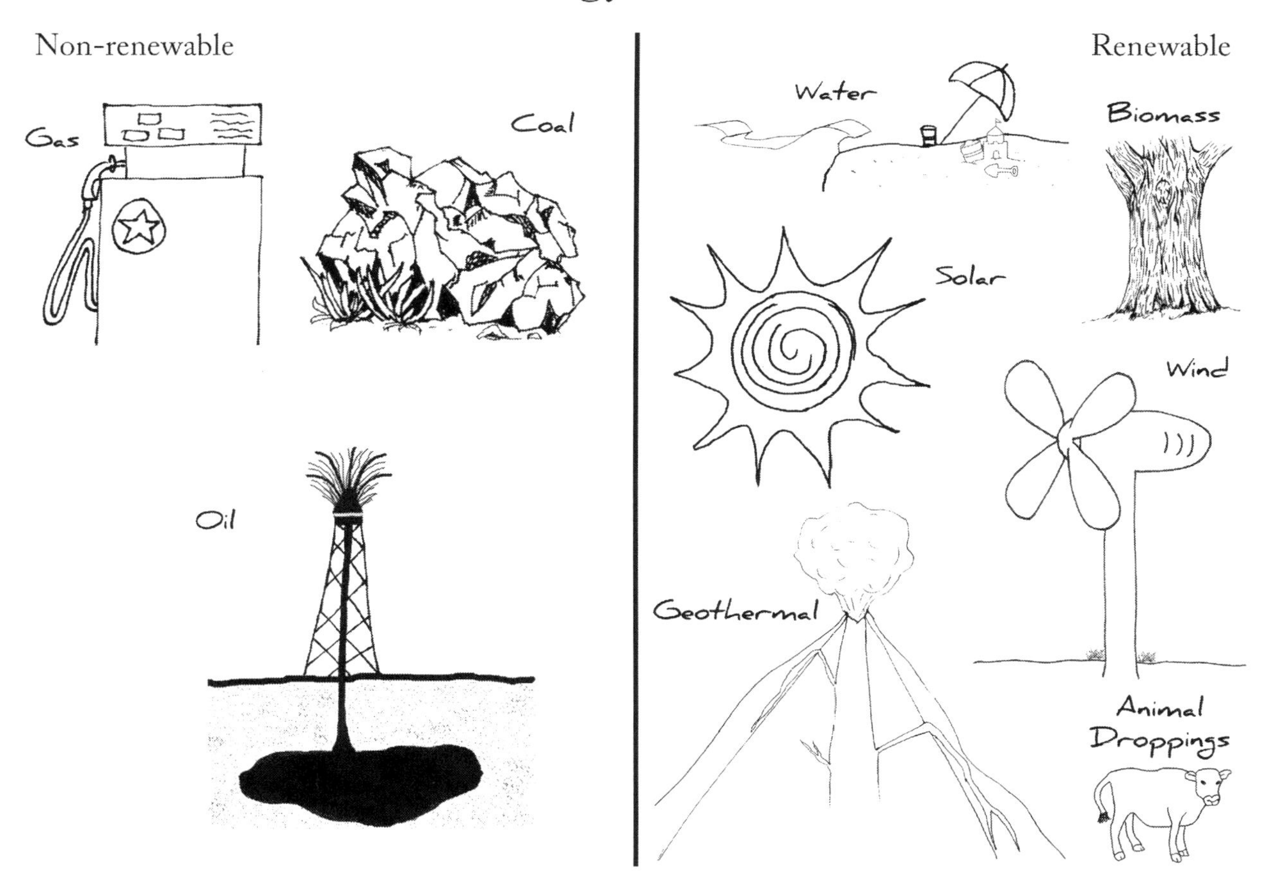

Student Assignment Sheet Week 8
Simple Machines

Experiment: Build a Simple Machine

Materials

✓ Materials will vary based on what you choose to build.

Procedure

1. Read the introduction to the experiment.
2. Choose one of the types of simple machines (lever, wheel and axle, gear, inclined plane, wedge, screw, or pulley) to build. (**Note**—*See the experiment sheet for an explanation of each of the types of simple machines.*)
3. Construct the machine and then test to see if it helps to perform work. For instance, if you choose to build a lever, test it by using a stack of books. First, pick up the stack on your own, then use the lever to pick up the stack. Which was easier to do?
4. Draw conclusions and complete the experiment sheet.

Vocabulary & Memory Work

- ☐ Vocabulary: fulcrum, input force, output force
- ☐ Memory Work—Continue to work on memorizing the different types of energy.

Sketch: Types of Simple Machines

- Label the following – lever, wheel and axle, inclined plane, wedge, screw, pulley, gear

Writing

- Reading Assignment: *DK Encyclopedia of Science* pp. 130-131 (Machines)
- Additional Research Readings
  - Ramps and Wedges: *KSE* pg. 300
  - Levers and Pulleys: *KSE* pg. 301
  - Wheels and Axles: *KSE* pg. 302
  - Machines: *UIDS* pp. 20-21

Dates

- c. 3rd century BC – Archimedes is said to have invented a screw pump to help get water from a reservoir source to the fields for irrigation.

Schedules for Week 8

Two Days a Week

Day 1	Day 2
☐ Do the "Build a Simple Machine" experiment, and then fill out the experiment sheet on SG pp. 66-67 ☐ Define fulcrum, input force, and output force on SG pg. 45 ☐ Enter the dates onto the date sheets on SG pp. 8-13	☐ Read pp. 130-131 from *DK EOS,* and then discuss what was read ☐ Color and label the "Types of Simple Machines" sketch on SG pg. 65 ☐ Prepare an outline or narrative summary; write it on SG pp. 68-69
Supplies I Need for the Week ✓ Materials will vary based on what you choose to build.	
Things I Need to Prepare	

Five Days a Week

Day 1	Day 2	Day 3	Day 4	Day 5
☐ Do the "Build a Simple Machine" experiment, and then fill out the experiment sheet on SG pp. 66-67 ☐ Enter the dates onto the date sheets on SG pp. 8-13	☐ Read pp. 130-131 from *DK EOS,* and then discuss what was read ☐ Write an outline on SG pg. 68	☐ Define fulcrum, input force, and output force on SG pg. 45 ☐ Color and label the "Types of Simple Machines" sketch on SG pg. 65	☐ Read one or all of the additional reading assignments ☐ Write a report on what you learned on SG pg. 69	☐ Complete one of the Want More Activities listed **OR** ☐ Study a scientist from the field of Physics
Supplies I Need for the Week ✓ Materials will vary based on what you choose to build.				
Things I Need to Prepare				

Additional Information Week 8

Experiment Information

- ☞ **Introduction –** (*from the Student Guide*) A simple machine is a tool that you can use to help you to do work. In other words, simple machines make the task of lifting or moving an object easier. There are seven main types of simple machines.
  - **Lever** – A rigid bar that is free to move around at a fixed point. Levers are frequently used to lift things, i.e. using a flat piece of metal to pry open a paint can, using a wheelbarrow to move dirt, or using a shovel to life a large rock.
  - **Wheel and axle** – Two disks or cylinders, each with a different radius. The wheel and axle simple machine is often used to turn or rotate things, i.e. using a screwdriver to turn a screw or using the steering wheel of a vehicle.
  - **Gears** – Toothed wheels that interlock in pairs; each one helps to drive the next. Gears are often used to consistently drive an object, like a watch or clock.
  - **Inclined plane** – A slanted surface that helps move objects up an incline. The inclined planes are typically ramps that can be used to move a heavy object from a lower level to a higher level.
  - **Wedge** – A v-shaped object whose sides are two inclined planes. Wedges, like knives, axes, and zippers, make it easier to separate two objects.
  - **Screw** – An inclined plane wrapped around a cylinder; this plane is also called the thread of the screw. These simple machines, including screws, nuts, and bolts, make it easier to drive an object into another object.
  - **Pulley** – A rope that fits in the groove of a wheel. Pulleys make it easier to pull a heavy load directly upward.

  In today's experiment, you are going to choose one of these types of simple machines to build.
- ☞ **Troubleshooting –** For ideas of simple machines the students can build easily at home, check out the following blog post:
  - 💻 http://elementalblogging.com/ideas-for-building-simple-machines/
- ☞ **Take it Further –** Have the students build a complex machine by using more than one type of simple machine.

Discussion Questions

1. What can machines do? (*Machines can take a small movement or force and turn it into a larger one. Machines can also change the direction of a force and apply it where it is needed.*)
2. What is a complex machine? (*A complex machine is a combination of simple machines connected together to accomplish a job.*)
3. What is the main benefit of using simple machines? (*Simple machines allow a small force to overcome a large one, making the overall job easier to do.*)
4. Name several examples of simple machines. (*Levers, pulleys, wedges, inclined planes, screws, wheels and axles, and gears are all considered simple machines.*)

Want More

- **Simple Machines Game –** Have the students play the simple machines game from the Museum of Science and Industry in Chicago.
  - http://www.msichicago.org/play/simplemachines/

Sketch Week 8

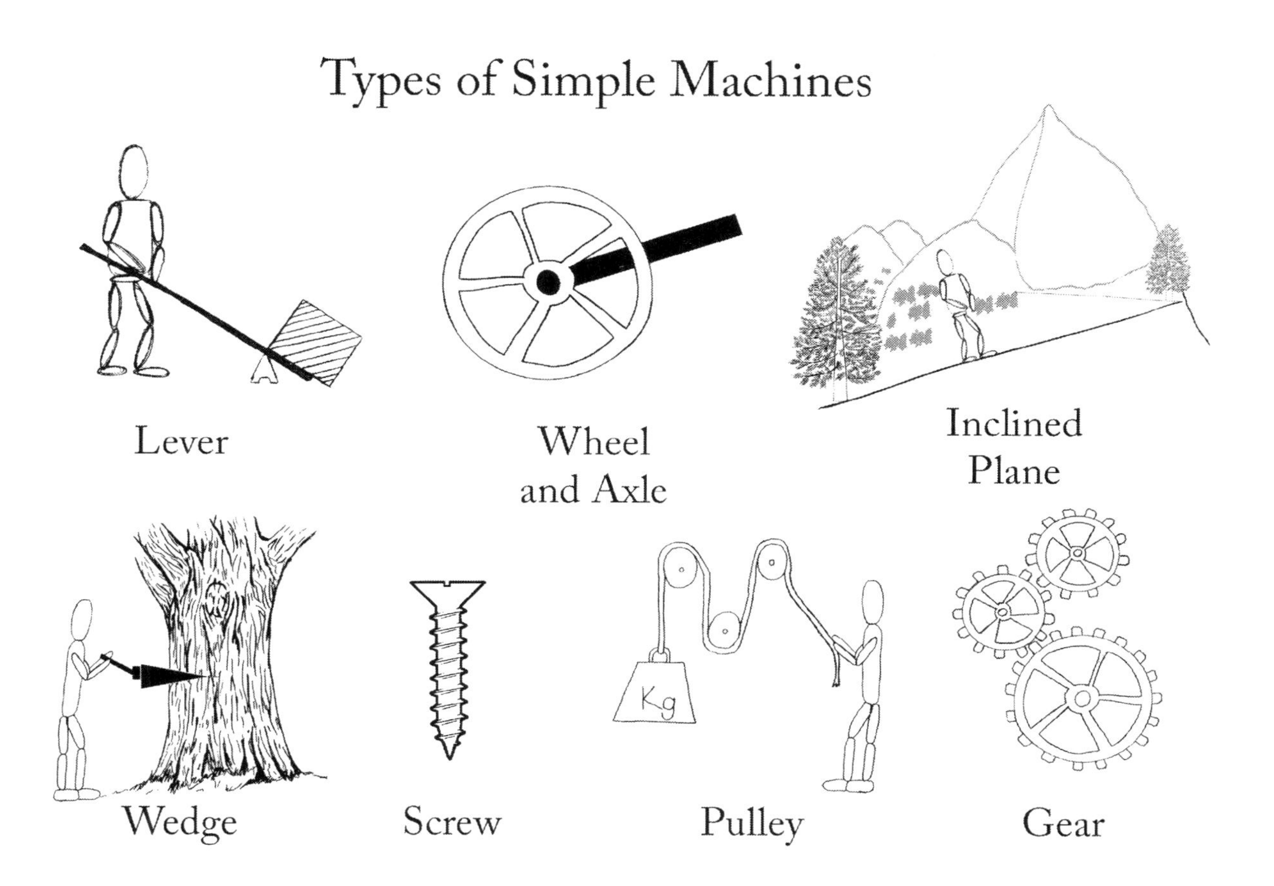

Unit 2 Energy
Unit Test Answers

Vocabulary Matching

1. D
2. H
3. F
4. B
5. A
6. J
7. K
8. I
9. E
10. C
11. G

True or False

1. True
2. False (*Energy is the ability to do work and work cannot be done without energy.*)
3. True
4. True
5. True
6. False (*Renewable energy sources will not run out, while non-renewable energy sources will eventually run out.*)
7. True
8. False (*Simple machines allow a small force to overcome a large one, making the overall job easier to do.*)

Short Answer

1. Kinetic energy is the energy of motion, but potential energy is the energy that an object has because of its position.
2. In a power plant, a furnace burns either coal or oil, which heats up water to form steam. The steam is then used to drive a turbine that is attached to a generator. The generator creates electricity, which can be sent to homes and buildings in the surrounding area.
3. If you spread force over a large area, you will reduce the amount of pressure exerted. If you concentrate the force over a small area, you will increase the amount of pressure exerted.
4. Levers, pulleys, wedges, inclined planes, screws, wheels and axles, and gears are all considered simple machines.
5. The students should have included four of the following types of energy in their answers.
   - ✓ **Mechanical energy** – The energy associated with the motion and position of an object. It is the sum of an object's kinetic and potential energy.
   - ✓ **Chemical energy** – The energy that is stored in chemical bonds. It is released when these bonds are broken in a chemical reaction.
   - ✓ **Thermal energy** – The energy that flows from one place to another due to changes in temperature. It is also known as heat.
   - ✓ **Electrical energy** – The energy associated with electrical charges.
   - ✓ **Electromagnetic energy** – The energy that travels through space in the form of waves, such as x-rays, light, and sound.
   - ✓ **Nuclear energy** – The energy stored in an atomic nucleus.

Unit 2 Energy
Unit Test

Vocabulary Matching

1. Energy ____
2. Kinetic energy ____
3. Potential energy ____
4. Work ____
5. Fluid ____
6. Pressure ____
7. Non-renewable energy resources ____
8. Renewable energy resources ____
9. Fulcrum ____
10. Input force ____
11. Output force ____

A. A substance that assumes the shape of the container it is in, such as a liquid or gas.

B. The transfer of energy that occurs when a force moves or changes an object.

C. The force you put into a machine.

D. The ability to do work.

E. The point on which a lever rests or is supported and on which it pivots.

F. The energy that an object has stored; it depends upon the object's weight and height.

G. The force that a machine exerts on an object.

H. The energy of an object in motion; it depends upon the object's mass and speed.

I. Energy sources that can be replaced in a relatively short period of time, such as wind and solar.

J. The amount of force pushing on a giving area.

K. Energy sources that exist in limited quantities, such as oil and coal.

True or False

1. ____________________ Several common forms of energy include chemical, heat, light, sound, nuclear, and electrical.

2. ____________________ Work and energy are not related in any way.

3. ____________________ Air pressure decreases as the altitude increases.

4. ____________________ Water pressure increases the deeper you go under water.

5. ____________________ Wind and water can be used to turn turbines that drive generators to create electricity.

6. ____________________ Non-renewable energy sources will not run out, while renewable energy sources will eventually run out.

7. ____________________ A complex machine is a combination of simple machines connected together to accomplish a job.

8. ____________________ Simple machines allow a large force to overcome a smaller one, making the overall job more difficult to do.

Short Answer

1. What is the difference between kinetic and potential energy?

2. How does a power plant work?

3. How are pressure, force, and area related?

4. What are the seven types of simple machines?

5. Name four different types of energy and explain what they are.

Physics Unit 3

Thermodynamics

Unit 3 Thermodynamics
Overview of Study

Sequence of Study

Week 9: Energy Conversion
Week 10: Heat
Week 11: Thermodynamics
Week 12: Engines

Materials by Week

Week	Materials
9-12	*Science Fair Project supplies will vary depending on the project the students choose to do.*

Vocabulary for the Unit

1. **Energy conversion** – The process of changing one form of energy into another.
2. **Entropy** – The degree of disorder in a given system.
3. **Heat** – A form of energy that flows from a place of high temperature to a place of lower temperature.
4. **Temperature** – A measure of how much heat energy is present in a substance.
5. **Absolute zero** – Theoretically, the lowest possible temperature or the point at which molecular motion virtually ceases to exist. (0 °K or -465.67 °F or – 273.15 °C)
6. **Conduction** – The movement of electricity or heat through a substance.
7. **Convection** – The movement within a fluid that is caused by the tendency of the hotter material to rise and the cooler material to sink, which results from the transfer of heat.
8. **Radiation** – The heat energy emitted by a solid object.
9. **Internal combustion engine** – An engine that generates power for movement through the burning of a fuel, such as coal, gas, or oil, and air.
10. **Power** – The rate at which work is done or energy is used.
11. **Steam engine** – An engine that generates power for movement through the use of steam, which rapidly expands and condenses.

Memory Work for the Unit

Laws of Thermodynamics

1. **Zeroth** – When two systems are in equilibrium with a third system, they are said to be in thermal equilibrium with each other.
2. **First** – Energy cannot be created or destroyed (also known as the Law of Conservation of Energy).
3. **Second** – Disorder (entropy) in the universe is always increasing.
4. **Third** – There is a theoretical point at which all molecular movement stops, which is

known as absolute zero.

Equation

- Celsius to Fahrenheit Equation

  $^{o}F = 1.8 \cdot {^{o}C} + 32$

  "^{o}F" stands for the temperature in Fahrenheit.
  "^{o}C" stands for the temperature in Celsius.

- Specific Heat Equation

  $Q = m \cdot c \cdot \Delta T$

  "Q" stands for the heat that is absorbed by a material.
  "m" stands for mass.
  "c" stands for specific heat.
  "ΔT" stands for change in temperature.

- Power Equation

  $P = \frac{W}{t}$

  "P" stands for power (measured in watts).
  "W" stands for work.
  "t" stands for time.

Notes

Student Assignment Sheet Week 9
Energy Conversion

Science Fair Project

This week, you will complete step one and begin step two of your Science Fair Project. You will be choosing your topic, formulating a question and doing some research about that topic.

1. **Choose your topic –** You should choose a topic in the field of physics that interests you, such as rockets. Next, come up with several questions you have relating to that topic, (e.g., "How does the design of the rocket affect its flight?" or "What type of rocket flies the highest?"). Then, choose the one question you would like to answer and refine it (e.g., "How does the fin design affect the rocket's flight?").
2. **Do Some Research –** Now that you have a topic and a question for your project, it is time to learn more about your topic so that you can make an educated guess (hypothesis) about the answer to your question. For the question stated above, you would need to research topics like rockets, fin designs, and flight. Begin by looking up the topic in the references you have at home. Then, make a trip to the library to search for more on the topic. As you do your research, write any relevant facts you have learned on index cards and be sure to record the sources you use.

Vocabulary & Memory Work

- ☐ Vocabulary: energy conversion, entropy
- ☐ Memory Work—This week, begin working on memorizing the laws of thermodynamics. (*See Unit Overview Sheet for a complete listing.*)

Sketch: Energy Chain

- Label the following – sun, nuclear energy is changed into heat and light energy, plants, light is changed into chemical energy, animals, chemical energy is changed into kinetic energy used of activities like breathing and moving

Writing

- Reading Assignment: *DK Encyclopedia of Science* pp. 138-139 (Energy Conversion)
- Additional Research Reading
  - Thermodynamics: *KSE* pg. 258 (Introduction and section on the First Law)

Dates to Enter

- c.550 BC – Ancient philosopher, Thales Miletus, believes that there is a conservation of some sort of hidden substance of which everything is made. (**Note—***Today we call this hidden substance mass energy.*)
- 1840's – James Joule does a number of experiments that lead to the development of the law of conservation of energy.
- 1850 – German scientist, Rudolf Clausius, suggests that the law of conservation of energy should be called the first law of thermodynamics.
- 1865 – Rudolf Clausius coins the term "entropy," which refers to unusable energy.
- 1920 – Ralph Fowler develops the zeroth law of thermodynamics.

Schedules for Week 9

Two Days a Week

Day 1	Day 2
☐ Decide on a Topic for the Science Fair Project and record on SG pg. 76 ☐ Do Some Research for the Science Fair Project and record on SG pp. 77 ☐ Enter the dates onto the date sheets on SG pp. 8-13	☐ Define energy conversion and entropy on SG pg. 72 ☐ Read pp. 138-139 from the *DK Encyclopedia of Science*, then discuss what was read ☐ Color and label the "Energy Chain" sketch on SG pg. 75 ☐ Prepare an outline or narrative summary, write it on SG pp. 78-79

Supplies I Need for the Week

✓ Index Cards

Things I Need to Prepare

Five Days a Week

Day 1	Day 2	Day 3	Day 4	Day 5
☐ Read pp. 138-139 from the *DK Encyclopedia of Science*, then discuss what was read ☐ Prepare an outline or narrative summary, write it on SG pp. 78-79	☐ Color and label the "Energy Chain" sketch on SG pg. 75 ☐ Define energy conversion and entropy on SG pg. 72	☐ Decide on a Topic for the Science Fair Project and record on SG pg. 76 ☐ Enter the dates onto the date sheets on SG pp. 8-13	☐ Do Some Research for the Science Fair Project and record on SG pg. 77	☐ Continue to research for the Science Fair Project and record on SG pp. 77

Supplies I Need for the Week

✓ Index Cards

Things I Need to Prepare

Additional Information Week 9

Notes

- **Science Fair Project** – If you choose not to have the students do a Science Fair Project, have them complete one of the Want More activities instead.

Science Fair Project

- Step 1: Choose the Topic – The students will be choosing a topic for their science fair project this week. Have them choose a topic in the field of physics that interests them. You can get ideas for projects from *Janice VanCleave's A+ Science Fair Projects* and *Janice VanCleave's A+ Projects in Physics: Winning Experiments for Science Fairs and Extra Credit.*
    1. **Key 1 ~ Decide on an area of physics.** The students should choose an area that fascinates them, something in physics that they want to know more about. You will begin by leading the students to brainstorm about things in physics that interest them. Have them rank these areas by degree of interest and then choose one area on which to focus. If their area is too broad, you will want them to narrow it down a bit. You can do this by asking them what they find interesting about the particular field.
    2. **Key 2 ~ Develop several questions about the area of physics.** Once the students have determined their area, they need to develop several questions about their topic that they can answer with their project. Good questions begin with how, what, when, who, which, why, or where. At this point, you are just getting them to think of possible questions.
    3. **Key 3 ~ Choose a question to be the topic.** Now that the students have several options of questions that they can answer with their science fair project, you will need to have them choose one of those questions for their project. Some of their questions will be easy to develop into an experiment for their science fair project that will determine the answer, but some will not.
- Step 2: Do Some Research – The students will also begin researching their topics. You may need to walk them through this process if they have not had much experience with doing research prior to this.
    1. **Key 1 ~ Brainstorm for research categories.** This is an important key, because developing relevant research categories before they begin to search for information will help the students to maintain a more focused approach. It will also help the students know where to begin their research and how to determine what information is important to their project and what is not. Keep in mind that some students may have a harder time coming up with categories that relate to their topic, so you may need to give them additional assistance. The students should have at least three categories and no more than five. This will help them to obtain relevant information as well as make it easier for them to write their report. Once the students have chosen their research categories, have them assign each category a number.
    2. **Key 2 ~ Research the categories.** Depending upon the students' experience with research, you may or may not have to walk them through this entire process. Either way, have them begin by looking at the reference material that they have close at hand, such as encyclopedias that they own or that are in the classroom. Then, they can look to their local library or the Internet for additional information. As the students uncover bits of relevant data, have them write each fact in their own words on a separate index card. They should number each card

at the top left with the category in which it fits, which will make the cards easier to organize. We also recommend that they assign a letter for each reference they use, which they can write in the right-hand top corner of each card. This way, after they organize and sort their cards, they will know which references they need to include in their bibliography. See the following article for more information on the index card system:

- http://elementalblogging.com/the-index-card-system/

Discussion Questions

1. Name one example of energy changing from one form to another. (*Students' answers will vary. One example would be a lightning strike, where electrical energy is converted into sound, light, and heat energy.*)
2. What is an energy chain? (*An energy chain shows the process of energy conversions.*)
3. What always happens when energy is converted? (*When energy is converted, some waste heat is always produced.*)
4. What is a perpetual motion machine? (*A perpetual motion machine is a machine that will keep working forever without an energy source.*) Why doesn't one exist? (*There are no perpetual motion machines because they always need more energy than they can give out due to waste heat energy. So, a constant source of energy is needed.*)

Want More

- **Testing the First Law of Thermodynamics** – You will need a large marble, a box-shaped container, a tape measure, and flour. Begin by filling the box-shaped container with a 2 in (5 cm) thick layer of flour. Have the students drop the marble from a height of 10 in (25 cm). Then, have them measure the width and depth of the crater. After they gather their measurements, have them shake the container to smooth out the surface of the flour. Have them repeat the process, dropping the marble from 20 in (50 cm) and 30 in (75 cm). Then, have the students draw some conclusions about what they have seen. (*The students should see that the higher the height the marble was drop from, the larger the impact crater. This is because the higher the marble, the greater amount of potential energy it has. That greater potential energy then translates into greater kinetic energy on impact, which results in a larger impact crater.*)

Sketch Week 9

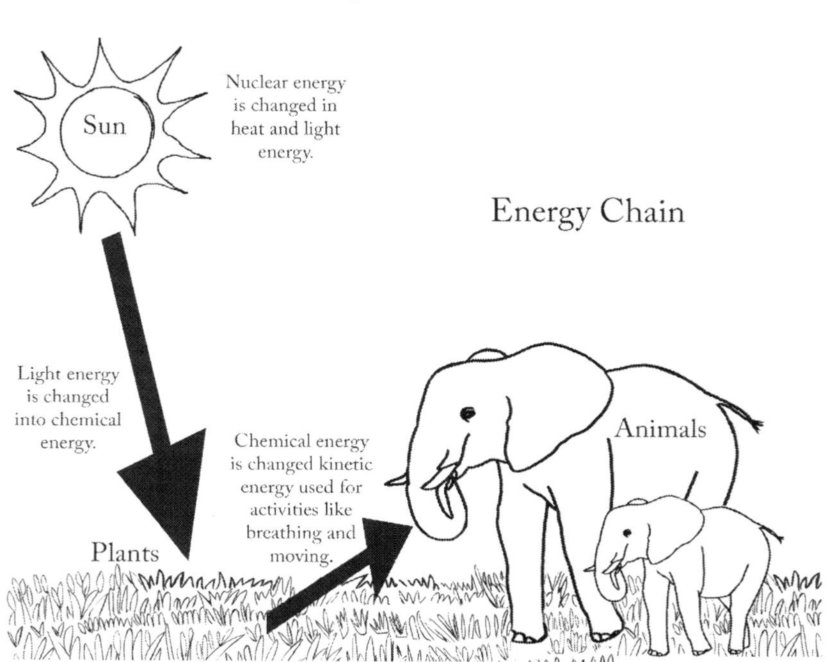

Student Assignment Sheet Week 10
Heat

Science Fair Project

This week, you will complete steps two through four of your Science Fair Project. You will be finishing your research, formulating your hypothesis, and designing your experiment.

2. **Do Some Research** – This week, you will finish your research. Then, organize your research index cards and write a brief report on what you have found out.
3. **Formulate a Hypothesis** – A hypothesis is an educated guess. For this step, you need to review your research and make an educated guess about the answer to your question. A hypothesis for the question asked in step one would be, "The rounder and smaller the fins, the faster the rocket will go."
4. **Design an Experiment** – Your experiment will test the answer to your question. It needs to have a control and several test groups. Your control will have nothing changed, while your test groups will change only one factor at a time. An experiment to test the hypothesis given above would be make several different designs fins for your rocket – one that is rounded and small, one that is rounded and normal size, one that is rounded and large, one that is squared off and small, one that squared off and normal size, one that is squared off and large, plus the standard fins that come with the rocket. If time allows, you can go ahead and begin your experiment this week.

Vocabulary & Memory Work

- ☐ Vocabulary: heat, temperature, absolute zero
- ☐ Memory Work—This week, continue to work on memorizing the laws of thermodynamics and the Celsius to Fahrenheit equation (°F = 1.8 • °C + 32).

Sketch

- There is no sketch this week, to allow for more time to work on the science fair project.

Writing

- Reading Assignment: *DK Encyclopedia of Science* pp. 140-141 (Heat)
- Additional Research Reading
  - Temperature: *UIDS* pp. 26-27
  - Thermodynamics: *KSE* pp. 258-259 (Read the remaining sections that you did not read last week.)

Dates to Enter

- 1714 – German physicist, Gabriel Fahrenheit, proposes a temperature scale where the lowest point was where he could cool brine and highest point was average internal temperature of the human body. It is eventually adopted and named after him.
- 1742 – Swedish physicist, Anders Celsius, develops a temperature scale where 0 represents the freezing point of water and 100 represents the boiling point of water. It is eventually adopted and named after him.
- 1848 – William Thomson, Lord Kelvin, develops an absolute temperature scale, known as the Kelvin scale.

Schedules for Week 10

Two Days a Week

Day 1	Day 2
☐ Finish your research, formulate a hypothesis for the Science Fair Project, and record on SG pg. 81 ☐ Design your experiment for your Science Fair Project, see SG pp. 81-82 for details ☐ Enter the dates onto the date sheets on SG pp. 8-13	☐ Define heat, temperature, and absolute zero on SG pg. 72 ☐ Read pp. 140-141 from the *DK Encyclopedia of Science*, then discuss what was read ☐ Prepare an outline or narrative summary, write it on SG pp. 84-85 ☐ Write the rough draft of the research report on SG pg. 83
Supplies I Need for the Week	
Things I Need to Prepare	

Five Days a Week

Day 1	Day 2	Day 3	Day 4	Day 5
☐ Finish your research, formulate a hypothesis for the Science Fair Project, and record on SG pg. 81	☐ Design your experiment for your Science Fair Project, see SG pp. 81-82 for details ☐ Write the rough draft of the research report on SG pg. 83	☐ Read pp. 140-141 from the *DK Encyclopedia of Science*, then discuss what was read ☐ Write an outline or list of facts on SG pg. 84	☐ Define heat, temperature, and absolute zero on SG pg. 72 ☐ Enter the dates onto the date sheets on SG pp. 8-13	☐ Read one or all of the additional reading assignments ☐ Prepare a report, write the report on SG pg. 85
Supplies I Need for the Week				
Things I Need to Prepare				

Additional Information Week 10

Notes

- **Absolute Zero** – Absolute zero was long thought of as the lowest possible temperature, but within the last few years, scientists have actually been able to forces gases below that temperature. The research is still ongoing and as it develops, changes to the third law of thermodynamics could be made. Here are several links to articles on the subject that you can use to discuss the matter with your students:
  - http://phys.org/news/2013-01-gas-temperature-absolute.html
  - http://www.livescience.com/25959-atoms-colder-than-absolute-zero.html
  - http://www.iflscience.com/physics/journey-other-side-absolute-zero
- **Science Fair Project** – If you choose not to have the students do a Science Fair Project, have them complete the Want More activity instead.

Science Fair Project

- *Step 2: Do Some Research* – This week, the students will finish their research and write a brief report for their project board.
  1. **Key 3 ~ Organize the information.** Once the students have finished their research, you need to have them organize and sort through the information that they have found. Begin by having the students sort their cards into piles using the research categories in the top left hand of their index card. Then, have the students read through each fact and determine five to seven of the most relevant pieces of information from each pile. You will need to help them as they decide which facts are relevant to their project (i.e., useful for answering their topical question) and which ones are not. These facts will then form the basis of their report.
  2. **Key 4 ~ Write a brief report.** Have the students determine the order they want to share their research categories. Normally, they would go from broad information about their subject to more specific information for their project. After they do this, they need to take the five to seven facts from the first category and turn them into a three to four sentence paragraph by combining the facts into a coherent passage. They will repeat this process until they have a three to five paragraph paper. Then, the students will need to edit and revise their paper so that it becomes a cohesive report. Finally, they will need to add in a bibliography with the resources they used for their report.
- *Step 3: Formulate a Hypothesis* – The students need to review their research and apply what they have learned to help them determine the answer to their question.
  1. **Key 1 ~ Review the research.** The students will need to review their research, so that it is fresh in their minds. You can do this by having them read over their report, or by having them read over each of the index cards they made. The level of involvement for this key will depend on how much time goes by between step two and step three of the science fair project.
  2. **Key 2 ~ Formulate an answer.** After the students have reviewed their research, they should also read over their question one more time. Once all the information is fresh in their minds, they are ready to make an educated guess at the answer to their question. Guide them to craft a response in the form of an if-then statement that they will be able to design an experiment to test. However, keep in mind that not all questions can be answered easily with if-then statements. As they design their experiment in the next step, you can still have them make a few adjustments to their hypotheses, if necessary.

Step 4: Design an Experiment – The students will also design their experiment this week. You may need to walk them through this process by suggesting ways they could test for the answer to their question.

1. **Key 1 ~ Choose a test.** You can ask the students what kind of a test could they use that would answer their question and prove their hypothesis either true or false. Have them write down each idea they have, but keep in mind that the students may need a fair amount of help with this process. If they find that they cannot come up with any options for testing their hypotheses, they may need to tweak their statements a bit. If they decide to do this, make sure they verify that the new versions still answer their original topical questions. Once the students have written down several ideas, have them review the options and choose one of the ideas for their projects.
2. **Key 2 ~ Determine the variables.** Now that the students have chosen a method to test their hypotheses, they need to determine the variables that will exist in their test. Here's a link to an article to help you understand the different types of variables:
   http://elementalblogging.com/experiment-variables/
   Have them answer the questions found in the student guide to determine their independent, dependent, and controlled variables.
3. **Key 3 ~ Plan the experiment.** Now that the students understand the variables that are at work, they are ready to use this information, along with their testing idea, to create an experiment design. You need to explain to them that they must have a control group as well as several test groups. The control group will have nothing changed, while each of the test groups will have only one change to the independent variable. The students should also plan on having several samples in each of their test groups. Once you have explained to the students the parameters of their experiment, they can begin formulating a plan by determining what their test groups will be. Then, they need to decide how long they have to run their tests. Once they have this information, they will write out their experiment design.

Discussion Questions

1. What are the three temperature scales? (*The three temperature scales are Fahrenheit, Celsius, and Kelvin.*)
2. What is the difference between heat and temperature? (*Heat is the energy that an object has because the molecules are moving. Temperature is the measure of how fast the object's molecules are moving, or the amount of heat energy the object has.*)
3. What happens to most objects as they are heated and cooled? (*When an object is heated, it typically expands. When an object is cooled, it typically shrinks.*)
4. What is latent heat? (*Latent heat is also known as hidden heat. It is the heat energy used to change the state of a material.*) And what happens to it as a material changes state? (*Latent heat is released as a liquid freezes and it is absorbed as a solid melts.*)

Want More

- **Expanding Gas –** Have the students try to reproduce the results shown on the top of pg. 141 (*Expanding Gases*) of the *DK Encyclopedia of Science*. You will need a glass bottle, a cork that fits the opening, hollow tube, and water.
- **Converting Temperatures** – Have the students use the worksheet on Appendix pg. 255 to practice converting temperatures from Celsius to Fahrenheit.
  Answers – 89.6 °F; 60.8 °F; 132.8 °F; 149 °C; 0 °C.

Student Assignment Sheet Week 11
Heat Transfer

Science Fair Project

This week, you will complete steps five and six of your Science Fair Project. You will carry out the experiment and record your observations and results.

5. **Perform the Experiment** – This week, you will perform the experiment you designed last week. Be sure to take pictures along the way as well as record your observations and results. (**Note—***Observations are a record of the things you see happening in your experiment. For instance, an observation would be that the rocket with the wide, rounded fins wobbled quite a bit during flight. Results are specific and measurable. For instance, results would be that the rocket with wide, rounded fins flew 50 feet into the sky and landed 65 feet from the launch stand. Observations are generally recorded in journal form, while results can be compiled into tables, charts, and graphs or relayed in paragraph form.*)
6. **Analyze the Data** – Once you have compiled your observations and results, you can use them to answer your question. You need to look for trends in your data and make conclusions from that. A possible conclusion to the rocket-fin design experiment would be, "I found that the smaller and more rounded the fins, the further and smoother the rocket will fly." If your hypothesis does not match your conclusion or your were not able to answer your question using the results from your experiment, you may need to go back and do some additional experimentation.

Vocabulary & Memory Work

- ☐ Vocabulary: conduction, convection, radiation
- ☐ Memory Work—This week, continue to work on memorizing the laws of thermodynamics and begin working on the specific heat equation.

Sketch

- There is no sketch this week, to allow for more time to work on the science fair project.

Writing

- Reading Assignment: *DK Encyclopedia of Science* pg. 142 (Heat Transfer)
- Additional Research Reading
  - Heat Transfer: *KSE* pp. 250-251
  - Transfer of Heat: *UIDS* pp. 28-29
  - Effects of Heat Transfer: *UIDS* pp. 30-31

Dates to Enter

- 1852 – William Thomson (Lord Kelvin) comes up with the idea of a "heat pump," a device that moves heat from a cold place to a hot one.
- 1892 – James Dewar invents the vacuum flask, which is designed to prevent the transfer of heat.

Schedules for Week 11

Two Days a Week

Day 1	Day 2
☐ Perform the experiment for your Science Fair Project and record your observations and results on SG pp. 87 and 89 ☐ Analyze your observations and results on SG pp. 87-88 ☐ Enter the dates onto the date sheets on SG pp. 8-13	☐ Define conduction, convection, and radiation on SG pg. 72 ☐ Read pg. 142 from the *DK Encyclopedia of Science*, then discuss what was read ☐ Prepare an outline or narrative summary, write it on SG pp. 90-91
Supplies I Need for the Week	
Things I Need to Prepare	

Five Days a Week

Day 1	Day 2	Day 3	Day 4	Day 5
☐ Perform the experiment for your Science Fair Project and record your observations and results on SG pp. 87 and 89	☐ Read pg. 142 from the *DK Encyclopedia of Science*, then discuss what was read ☐ Write an outline or list of facts on SG pg. 90	☐ Define conduction, convection, and radiation on SG pg. 72 ☐ Enter the dates onto the date sheets on SG pp. 8-13	☐ Read one or all of the additional reading assignments ☐ Prepare a report, write the report on SG pg. 91	☐ Finish the experiment for your Science Fair Project ☐ Analyze your observations and results on SG pp. 87-88
Supplies I Need for the Week				
Things I Need to Prepare				

Additional Information Week 11

Notes

- **Science Fair Project** – If you choose not to have the students do a Science Fair Project, have them complete the Want More activity instead.

Science Fair Project

- *Step 5: Perform the Experiment* – The students will be performing the experiment and recording their observations and results. Be sure to check in with them to see how they are doing.
    1. **Key 1 ~ Get ready for the experiment.** The students already have a plan in place, but there are still a few things they need to do before beginning their experiment. They need to look at a calendar and make sure that they will be home for the duration of the trial because they will need to be there to make observations and record results on each day of testing. The students also need to gather and prep any materials that they will be using during their experiment.
    2. **Key 2 ~ Run the experiment.** The students have done a lot of work to reach this point, but that preparation has paved a smooth road for their experiment. At this point, they are familiar with their research and their design, so they should be able to carry out their testing with little to no help. You want to make sure that the students write down a list of things they need to check each day during the experiment. Be sure that they include taking pictures of what they see on their list as they will need these images for their project board.
    3. **Key 3 ~ Record any observations and results.** As the students run their experiment, they need to compile their observations and results. Observations are the record of the things the scientist sees happening in an experiment, while results are specific and measurable. Observations are generally recorded in journal form, while results can be compiled into tables, charts, or graphs. You will need to help the students create a table to record their results, as well as provide them with a journal for their observations. Once they finish their experiment, you may need to help them chart or graph their data.
- *Step 6: Analyze the Data* – The students will now analyze their observations and results to draw conclusions from their experiment.
    1. **Key 1 ~ Review and organize the data.** The students need to analyze their observations and results to determine if their hypotheses are true or false. To do this, you need to have them read over each of their journal entries and note any trends in their observations. You also need to have the students interpret the charts or graphs they created in the last step and write down the information that they can glean from them.
    2. **Key 2 ~ State the answer.** Now that the students have noted trends from their observations and interpreted information from their results, they can use those data to answer their question. They need to first determine if they have proved their hypotheses true or false. Once the students have decided if their hypotheses statements were true or false, they can craft a one sentence answer to their original topical questions from step one. Their statements should begin with, "I found that ___" or "I discovered that ___." In the rare case that the students are unable to state an answer to their question, they need to take what they have learned, go back to the drawing table, and redesign their experiment.

3. **Key 3 ~ Draw several conclusions.** When the students draw conclusions, they are putting into words what they have learned from their project. Each conclusion should include the following information:
   - ☑ The answer to student's question;
   - ☑ Whether or not the student's hypothesis was proven true (***Note:*** *If the hypothesis was proven false, the student should state why.*);
   - ☑ Any problems or difficulties the student ran into while performing the experiment;
   - ☑ Anything interesting the student discovered that he or she would like to share;
   - ☑ Ways that the student would like to expand the experiment in the future.

   It should be one paragraph, or about four to six sentences in length. Have the students begin their concluding paragraph with the statement they wrote for the previous key.

Discussion Questions

1. How does heat travel? (*Heat always travels from something hot to something cold.*)
2. What are the three ways heat can travel? Give a brief description of each. (*Heat can travel by conduction, convection, and radiation. Conduction is how heat travels through a solids. Convection is how heat travels through liquids and gases. Radiation is how heat travels through empty spaces.*)
3. What is the purpose of insulation? (*Insulation helps to prevent heat loss.*)
4. How does a vacuum flask prevent the three types of heat travel? (*A vacuum flask prevents conduction and convection with a vacuum between its double walls. It prevents radiation by silvering the walls.*)

Want More

- **Specific Heat Worksheet** – Have the students practice calculating specific heat with the worksheet on Appendix pg. 256.
  Answers – 39,188 J; 27,672 J; 9.4 °C.
- **Convection Experiment** – Have the students explore convection of heat. They will need three glasses, red and blue food coloring, ice cold water, and hot water (**Note**—*Warn the students not to handle the hot water without the proper protection.*). They should begin by filling one of the glasses two thirds of the way with cold water and add three drops of blue food coloring. Have the students observe what happens and record the time it takes for the food coloring to completely mix. Next have the students fill the other glass two thirds of the way with hot water and add two drops of food coloring. Have them observe what happens and record the time it takes for the food coloring to completely mix. Then, have the students pour one quarter of the hot water very slowly into the cold water, observe what happens, and record it. After 30 minutes, have the students check the glass once more to observe and record what has happened. (*The students should see that it took about three times as long for the food coloring to fully mix into the cold water than in the hot water. They should also see that the hot red water layer "floats" on top of the cold blue water layer . After 30 minutes, they should see that the water is now purple because the two layers have mixed. This is because hot water molecules move faster than cold water molecules which causes the food coloring in the hot water to mix in much quicker. Also hot water molecules are more spread out, making the same volume of hot water to be less dense than cold water. This allows the hot water to appear to float on the surface of the cold water. As the hot water cools and the cool water warms up, the two mix turning the water purple.*)

Student Assignment Sheet Week 12
Engines

Science Fair Project

This week, you will complete steps seven and eight of your Science Fair Project. You will be writing and preparing a presentation of your Science Fair Project.

7. **Create a Board** – This week, you will be creating a visual representation of your science fair project that will serve as the centerpiece of your presentation. You will begin by planning the look of your board, then move onto preparing the information, and finally you will pull it all together.
8. **Give a Presentation** – After you have completed your presentation board, determine if you would like to include part of your experiment in your presentation. Then, prepare a five minute talk about your project, be sure to include the question you tried to answer, your hypothesis, a brief explanation of your experiment and the results plus the conclusion to your project. Be sure to arrive on time for your presentation. Set up your project board and any other additional materials. Give your talk and then ask if there are any questions. Answer the questions and end your time by thanking whoever has come to listen to your presentation.

Vocabulary & Memory Work

- ❒ Vocabulary: internal combustion engine, power, steam engine
- ❒ Memory Work—This week, continue to work on memorizing the laws of thermodynamics and the power equation.
  - Power (P) = Work (W) / Time (t)

Sketch

- There is no sketch this week, to allow for more time to work on the science fair project.

Writing

- Reading Assignment: *DK Encyclopedia of Science* pp. 143-144 (Engines)
- Additional Research Reading
  - Combustion: *KSE* pp. 254

Dates to Enter

- 1712 – The first steam engine is built by Thomas Newcomen.
- 1765 – James Watt improves upon the original Newcomen steam engine.
- 1860 – The first internal combustion engine is built by Etienne Lenoir. He uses coal gas and air for fuel.
- 1884 – The first steam engine to generate electricity is invented by Charles Parsons.
- 1926 – The first rocket propelled by liquid fuel is launched by Robert Goddard.

Schedules for Week 12

Two Days a Week

Day 1	Day 2
☐ Define internal combustion engine and steam engine on SG pp. 73 ☐ Read pp. 143-144 from the *DK Encyclopedia of Science*, then discuss what was read ☐ Prepare an outline or narrative summary, write it on SG pp. 96-97	☐ Prepare your science fair project board and present your Science Fair Project, see SG pp. 94-96 for details ☐ Enter the dates onto the date sheets on SG pp. 8-13 ☐ Take the Unit 3 test
Supplies I Need for the Week	
Things I Need to Prepare	

Five Days a Week

Day 1	Day 2	Day 3	Day 4	Day 5
☐ Begin working on your science fair project board, see SG pp. 94-95 for details	☐ Read pp. 143-144 from the *DK Encyclopedia of Science*, then discuss what was read ☐ Write an outline or list of facts on SG pg. 96	☐ Define internal combustion engine and steam engine on SG pp. 73 ☐ Enter the dates onto the date sheets on SG pp. 8-13	☐ Read one or all of the additional reading assignments ☐ Prepare a short report, write the report on SG pg. 97	☐ Present the Science Fair Project, see SG pg. 96for details ☐ Take the Unit 3 Test
Supplies I Need for the Week				
Things I Need to Prepare				

Additional Information Week 12

Notes

- **Science Fair Project** – If you choose not to have the students do a Science Fair Project, have them complete one of the Want More activities instead.

Science Fair Project

- *Step 7: Create a Board* – In this step, the students will pull together all the information they have learned to create a presentation board.
  1. **Key 1 ~ Plan out the board.** The science fair project board is the visual representation of the students' hard work, so you definitely want them to put as much effort into this step as they have into the others. The board will have specific sections that are set, but they should personalize the look with color and graphics that suit their tastes and match their projects. Please see the Appendix pg. 246 for a more detailed explanation of the science fair project board layout.
  2. **Key 2 ~ Prepare the information.** The students have put in a lot of effort until this point, but the work they have done in the previous steps will make it easier for them to prepare the information for their board. The students need to type up the information and choose a font and font size for their board. Please see the Appendix pg. 246 for a more detailed explanation of what each section should include.
  3. **Key 3 ~ Put the board together.** Now that the students have planned out their science fair project boards and prepared the information, they are ready to pull it all together. They need to cut out the decorative elements and glue them to the backboards. Then, they need to print and cut out their informational paragraphs. For added depth, they can glue the paragraphs onto a foam board before adding the information. Finally, the students should add their titles and the finishing touches to their board.
- *Step 8: Give a Presentation* – This step gives the students a chance to communicate with an audience what they have learned from their project. The best way to achieve this is to have the students participate in a Science Fair where their projects will be judged, but if that's not possible, don't skip this key. The students can still present their projects to their families or to a group of their peers.
  1. **Key 1 ~ Prepare the presentation.** Once the students have finished their project boards, they can begin to work on their presentation. They should each prepare a brief five minute talk about their science fair project. This talk should include the question each tried to answer, their hypotheses, a brief explanation of their experiment, the results, and the conclusion to their project. You will need to guide the students as they each turn their information paragraphs into an outline for their presentation. This outline should highlight the main points that the students want to cover for their presentations.
  2. **Key 2 ~ Practice the presentation.** Once the students have finished preparing their outlines for their talks, have them practice in front of a mirror. They should practice looking at the audience while pointing to the different sections on their project board as they present. Once they feel confident with their presentations, have them give a practice talk to you. Be sure to give them feedback, so that they can make the necessary changes before they present their science fair projects to a group.

3. **Key 3 ~ Share the presentation.** It is important to have the students present their work to an audience and answer related questions from the group. This will reinforce what they have learned as well as help them to discern how to communicate what they know.

Discussion Questions

1. What does every engine do? (*Every engine converts the energy stored in a fuel into energy for movement.*)
2. What is the main difference between an internal combustion engine and a steam engine? (*In an internal combustion engine, the fuel is burned inside the engine. In a steam engine, the fuel is burned outside of the engine.*)
3. How does an internal combustion engine work? (*The piston descends to suck fuel into the chamber. Then, the piston ascends to compress the fuel. A spark ignites the fuel and explosion forces the piston down. Finally, the piston ascends to force out the burned fuel as exhaust. The constant motion of the piston provides energy for the movement of the object.*)
4. How does a steam engine work? (*The burning of a fuel heats up a tank of water, which makes steam. The steam expands, forcing the object to move.*)
5. How do rockets work? (*Rockets move forward in response to hot gases being forced out of the back end.*)
6. What is jet propulsion? (*Jet propulsion is the pushing forward of an object by the stream of a liquid or gas.*)

Want More

- **Power Worksheet** – Have the students practice calculating power with the worksheet on Appendix pg. 257.
  Answers:
  1. 125 Watts
  2. 77 Watts
  3. 2236 Watts
- **Balloon Engine** – Have the students make their own vehicles powered by balloons. You will need to provide each of them with Lego bricks, Lego wheels, and a balloon. Their designs will vary and you can have a race afterwards to see whose balloon-engine-vehicle design was the most efficient.

Unit 3: Thermodynamics
Unit Test Answers

Vocabulary

1. C
2. F
3. A
4. D
5. J
6. B
7. G
8. I
9. H
10. K
11. E

True or False

1. True
2. False (*When energy is converted, some waste heat is always produced.*)
3. False (*When an object is heated, it typically expands. When an object is cooled, it typically shrinks.*)
4. True
5. True
6. False (*Insulation helps to prevent heat loss.*)
7. True
8. True

Short Answer

1. Students' answers will vary. One example would be a lightning strike, where electrical energy is converted into sound, light, and heat energy.
2. Heat is the energy that an object has because the molecules are moving. Temperature is the measure of how fast the object's molecules are moving, or the amount of heat energy the object has.
3. Heat can travel by conduction, convection, and radiation. Conduction is how heat travels through a solids. Convection is how heat travels through liquids and gases. Radiation is how heat travels through empty spaces.
4. In an internal combustion engine, the fuel is burned inside the engine. In a steam engine, the fuel is burned outside of the engine.
5. The laws are thermodynamics are:
   - **Zeroth –** When two systems are in equilibrium with a third system, they are said to be in thermal equilibrium with each other.
   - **First –** Energy cannot be created or destroyed (also known as the Law of Conservation of Energy).
   - **Second –** Disorder (entropy) in the universe is always increasing.
   - **Third –** There is a theoretical point at which all molecular movement stops, which is known as absolute zero.

Unit 3: Thermodynamics
Unit Test

Vocabulary Matching

1. Energy conversion ____
2. Entropy ____
3. Heat ____
4. Temperature ____
5. Absolute zero ____
6. Conduction ____
7. Convection ____
8. Radiation ____
9. Internal combustion engine ____
10. Power ____
11. Steam engine ____

A. A form of energy that flows from a place of high temperature to a place of lower temperature.

B. The movement of electricity or heat through a substance.

C. The process of changing one form of energy into another.

D. A measure of how much heat energy is present in a substance.

E. An engine that generates power for movement through the use of steam, which rapidly expands and condenses.

F. The degree of disorder in a given system.

G. The movement within a fluid that is caused by the tendency of the hotter material to rise and the cooler material to sink, which results from the transfer of heat.

H. An engine that generates power for movement through the burning of a fuel, such as coal, gas, or oil, and air.

I. The heat energy emitted by a solid object.

J. Theoretically, lowest possible temperature or the point at which molecular motion virtually ceases to exist. (0 °K or -465.67 °F or – 273.15 °C)

K. The rate at which work is done or energy is used.

True or False

1. ____________________ An energy chain shows the process of energy conversions.
2. ____________________ When energy is converted, no waste heat is ever produced.
3. ____________________ When an object is heated, it typically shrinks. When an object is

cooled, it typically expands.

4. ____________________ The three temperature scales are Fahrenheit, Celsius, and Kelvin.

5. ____________________ Heat always travels from something hot to something cold.

6. ____________________ Insulation helps to increase the amount of heat loss.

7. ____________________ Every engine converts the energy stored in a fuel into energy for movement.

8. ____________________ Jet propulsion is the pushing forward of an object by the stream of a liquid or gas.

Short Answer

1. Give one example of energy changing from one form to another.

2. What is the difference between heat and temperature?

3. What are the three ways heat can travel? Give a brief description of each.

4. What is the main difference between an internal combustion engine and a steam engine?

5. Define the laws of thermodynamics.

Physics Unit 4

Sound

Unit 4 Sound
Overview of Study

Sequence of Study

Week 13: Sound
Week 14: Sound Waves
Week 15: Hearing Sound
Week 16: Acoustics

Materials by Week

Week	Materials
13	Glass bottle, Bell, Cork that fits the top of the glass bottle, Thread, Needle, Match
14	Shallow glass bowl or cup, Water, Music player
15	Plastic jar, or small flower pot, A piece of latex material large enough to cover the lid of your jar (like the kind used for exercise bands), 1" plastic tubing, Rubber band, Air-dry clay, Salt
16	Partner, Blindfold

Vocabulary for the Unit

1. **Mechanical wave** – A wave that travels through a medium, such as air, water, or solids.
2. **Sound** – A mechanical wave, or vibration, that travels through a medium, such as air, and can be heard when it reaches a person's or animal's ear.
3. **Vibration** – A quick back and forth movement, for example when sound waves causes a nearby glass of water to vibrate.
4. **Amplitude** – The size of a vibration or the height of a wave.
5. **Frequency** – The number of waves that pass a given point in a second.
6. **Wavelength** – The distance between the crest of one wave to the crest of another.
7. **Antinoise** – Produced when two sound waves overlap and cancel each other out.
8. **Resonate frequency** – The frequency at which an object naturally begins to vibrate.
9. **Acoustics** – The study of how sound travels in a given space.
10. **Echolocation** – The locating of objects through the use of reflected sound.

Memory Work for the Unit

Types of Mechanical Waves

1. **Transverse Wave** – A wave that causes the medium to vibrate perpendicular to the direction in which the wave travels.

2. **Longitudinal Waves** – A wave that causes the medium to vibrate parallel to the direction in which the wave travels.
3. **Surface Waves** – A wave that travels along the surface separating two media.

Equation

- Speed of Waves Equation

  $v = \lambda \bullet f$

  "v" stands for speed.
  "λ" stands for wavelength.
  "f" stands for frequency.

Notes

Student Assignment Sheet Week 13
Sound

Experiment: Does sound travel in a vacuum?

Materials

- ✓ Glass bottle
- ✓ Bell
- ✓ Cork that fits the top of the glass bottle
- ✓ Thread
- ✓ Needle
- ✓ Match

Procedure

1. Read the introduction to the experiment and answer the question for the hypothesis section.
2. Thread the needle and then pass the thread through the bell and up through the cork, so that it will hang inside the bottle and swing freely.
3. Lightly place the cork in top of the bottle and shake gently. Observe if you hear any noise and write your observations on your experiment sheet.
4. Next, remove the cork. Light a match and drop it into the bottle. Quickly replace the cork and make sure it is firmly in place. Set the bottle on the counter and wait for the match to burn out.
5. Once the bottle is cool to the touch, gently shake it and observe if you hear any noise. Write your observations on your experiment sheet.
6. Draw conclusions and complete the experiment sheet.

Vocabulary & Memory Work

- ☐ Vocabulary: mechanical wave, sound, vibration
- ☐ Memory Work—This week, begin working on memorizing the types of mechanical waves.
  - Transverse Wave – A wave that causes the medium to vibrate perpendicular to the direction in which the wave travels.
  - Longitudinal Waves – A wave that causes the medium to vibrate parallel to the direction in which the wave travels.
  - Surface Waves – A wave that travels along the surface separating two medias.

Sketch: Transverse and Longitudinal Waves

- Label the following – transverse wave, longitudinal wave.
- Add a double arrow to show the direction of the original movement.

Writing

- Reading Assignment: *DK Encyclopedia of Science* pp. 178-179 (Sound)
- Additional Research Readings
  - Waves: *UIDS* pp. 34-35
  - Sound as Changes of Pressure: *KSE* pp. 312-313

Dates

- 1708 – William Derham successfully establishes the speed of sound.
- 1890's – Ernst Mach describes how shock waves form and, along with his son Ludwig, develops a way to take pictures of the shadow of an invisible shock wave.

Schedules for Week 13

Two Days a Week

Day 1	Day 2
☐ Do the "Does sound travel in a vacuum?" experiment, and then fill out the experiment sheet on SG pp. 104-105 ☐ Define mechanical wave, sound, and vibration on SG pg. 100 ☐ Enter the dates onto the date sheets on SG pp. 8-13	☐ Read pp. 178-179 from *DK EOS,* and then discuss what was read ☐ Color and label the "Transverse and Longitudinal Waves" sketch on SG pg. 103 ☐ Prepare an outline or narrative summary; write it on SG pp. 106-107

Supplies I Need for the Week

✓ Glass bottle, Cork that fits the top of the glass bottle
✓ Bell, Thread, Needle
✓ Match

Things I Need to Prepare

Five Days a Week

Day 1	Day 2	Day 3	Day 4	Day 5
☐ Do the "Does sound travel in a vacuum?" experiment, and then fill out the experiment sheet on SG pp. 104-105 ☐ Enter the dates onto the date sheets on SG pp. 8-13	☐ Read pp. 178-179 from *DK EOS,* and then discuss what was read ☐ Write an outline on SG pg. 106	☐ Define mechanical wave, sound, and vibration on SG pg. 100 ☐ Color and label the "Transverse and Longitudinal Waves" sketch on SG pg. 103	☐ Read one or all of the additional reading assignments ☐ Write a report on what you learned on SG pg. 107	☐ Complete one of the Want More Activities listed **OR** ☐ Study a scientist from the field of Physics

Supplies I Need for the Week

✓ Glass bottle, Cork that fits the top of the glass bottle
✓ Bell, Thread, Needle
✓ Match

Things I Need to Prepare

Additional Information Week 13

Experiment Information

- ☞ **Introduction** – (*from the Student Guide*) Sound is a mechanical wave or vibration that can be heard. Both animals and humans have special organs, known as ears, which are able to detect the vibrations caused by sound waves. The human ear can hear sound when an object vibrates at sixteen times per second and as high as twenty thousand times per second. In today's experiment, you are going to test whether the human ear can detect sound when the object is in a vacuum.
- ☞ **Results** – The students should hear the bell ringing the first time, but they should hear nothing the second time.
- ☞ **Explanation** – Sound waves must travel through a medium, like air, water, or a solid surface. In the first case, there was still some air in the bottle, which is why the students could hear the bell ringing. In the second case, the match burned up the air within the bottle, creating a vacuum. Since there was no medium in which the sound waves could travel, no sound is heard.
- ☞ **Troubleshooting** – If the students still hear a sound for the first test, check two things:
  - The cork completely seals the top of the bottle.
  - The match burned out on its own before it ran out of fuel; in other words, before it reached the other end. If not, the students have air getting into the bottle somehow.
- ☞ **Take it Further** – Have the students fill the bottle completely with water. Then, have them shake the bottle to see if they can hear the bell ring. (*They should be able to hear the bell ring, but it will take a bit of shaking as the bell may tend float a bit.)*

  Note – You can use a wire to attach the bell to the cork in the water to prevent floating.

Discussion Questions

1. What is sound caused by? (*Sound is caused by the rapid motion of particles of matter colliding with one another, known as vibration. This causes energy to be passed along as a wave.*)
2. What are two ways that sound travels differently through water than air? (*When sound travels through the water, it loses less energy and moves faster than it does through air.*)
3. Why does sound travel more quickly through a solid than a gas? (*Sound travels more quickly through a solid because the molecules are closer together and more firmly in place. So, they bounce back into place quicker than the molecules in a gas.*)
4. What causes a sonic boom? (*A sonic boom is cause when the sound of a supersonic jet catches up to the speed of sound, causing a shock wave of sound.*)

Want More

- **Wave Simulation** – Have the student do the following on-line wave on a string simulation from PhET:
  - http://phet.colorado.edu/en/simulation/wave-on-a-string
- **Wave Test** – Have the students look at the three different types of waves.
  1. **Transverse wave** – Have two students hold a 5 foot ribbon out straight between them. Then, have one student begin to move their hand up and down several times. Observe the

wave that is created in the ribbon.

2. **Longitudinal wave** – Have two student hold either end of a slinky across a table. Then, have one student begin to push and pull the slinky towards and away from the other student several times. Observe the wave that is created in the slinky.
3. **Surface wave** – Have the students fill a large shallow bowl with water and set it on a flat surface. Once the water has settled, have them float a cork in the center. Then, have the students gently tap on the side of the bowl. Observe the wave that is created and how that wave affects the cork.

Sketch Week 13

Transverse and Longitudinal Waves

Longitudinal Waves

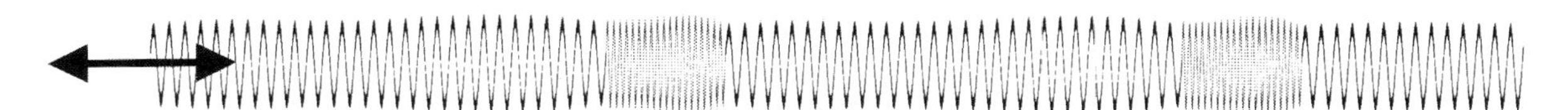

Original Movement

Transverse Waves

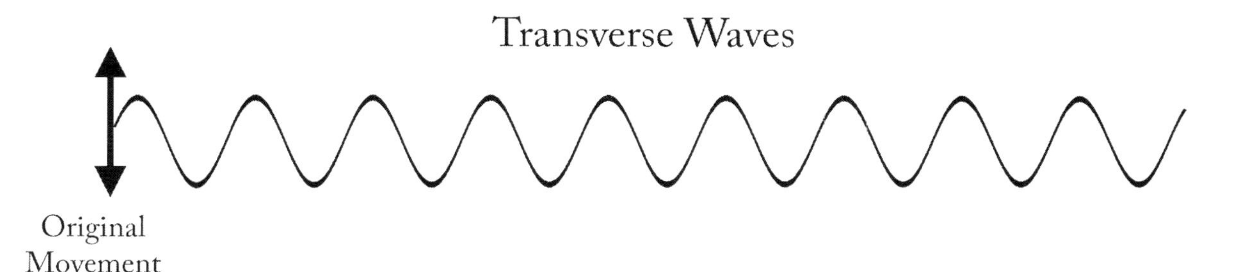

Student Assignment Sheet Week 14
Sound Waves

Experiment: Does distance affect the number of vibrations experience by a medium?

Materials
- ✓ Shallow glass bowl or cup
- ✓ Water
- ✓ Music Player
- ✓ Measuring tape

Procedure
1. Read the introduction to the experiment and answer the question for the hypothesis section.
2. Fill the bowl with water and set it right next to the speaker of the music player. Choose a fast selection of music and play it at a relatively loud volume. Observe the changes in the water in the bowl and sketch what you see.
3. Now, move the player 12 in (30 cm) away from the bowl. Play the same selection of music once more. Observe the changes in the water in the bowl and sketch what you see.
4. Then, move the player 24 in (60 cm) away from the bowl. Play the same selection of music again. Observe the changes in the water in the bowl and sketch what you see.
5. Finally, move the player 36 in (90 cm) away from the bowl. Play the same selection of music for the final time. Observe the changes in the water in the bowl and sketch what you see.
6. Draw conclusions and complete the experiment sheet.

Vocabulary & Memory Work
- ☐ Vocabulary: amplitude, frequency, wavelength
- ☐ Memory Work—This week, continue to work on memorizing the types of mechanical waves along with the speed of waves equation:
  - Speed (v) = Wavelength (λ) • Frequency (f)

Sketch: Sound Wave
- Label the following – crest, trough, wavelength, amplitude.

Writing
- Reading Assignment: *DK Encyclopedia of Science* pg. 180 (Measuring Sound), pg. 188 (Sound Recording)
- Additional Research Readings
  - Wave Motion: *KSE* pp. 314-315
  - Vibrations: *KSE* pp. 316-317
  - Sound Waves: *UIDS* pp. 40-41

Dates
- 1877 – The first sound recording is made by Thomas Edison on a phonograph through the mechanical vibrations of a needle.
- 1887 – Henrich Hertz proves that electricity can be transmitted in electromagnetic waves. He was also the first to produce and detect radio waves.

Schedules for Week 14

Two Days a Week

Day 1	Day 2
☐ Do the "Does distance affect the amount of vibrations experience by a medium?" experiment, and then fill out the experiment sheet on SG pp. 110-111 ☐ Define amplitude, frequency, and wavelength on SG pg. 100 ☐ Enter the dates onto the date sheets on SG pp. 8-13	☐ Read pp. 180 and 188 from *DK EOS,* and then discuss what was read ☐ Color and label the "Sound Wave" sketch on SG pg. 109 ☐ Prepare an outline or narrative summary; write it on SG pp. 112-113
Supplies I Need for the Week ✓ Shallow glass bowl or cup, Water ✓ Music Player ✓ Measuring tape	
Things I Need to Prepare	

Five Days a Week

Day 1	Day 2	Day 3	Day 4	Day 5
☐ Do the "Does distance affect the amount of vibrations experience by a medium?" experiment, and then fill out the experiment sheet on SG pp. 110-111 ☐ Enter the dates onto the date sheets on SG pp. 8-13	☐ Read pp. 180 and 188 from *DK EOS,* and then discuss what was read ☐ Write an outline on SG pg. 112	☐ Define amplitude, frequency, and wavelength on SG pg. 100 ☐ Color and label the "Sound Wave" sketch on SG pg. 109	☐ Read one or all of the additional reading assignments ☐ Write a report on what you learned on SG pg. 113	☐ Complete one of the Want More Activities listed **OR** ☐ Study a scientist from the field of Physics
Supplies I Need for the Week ✓ Shallow glass bowl or cup, Water ✓ Music Player ✓ Measuring tape				
Things I Need to Prepare				

Additional Information Week 14

Experiment Information

- ☞ **Introduction** – (*from the Student Guide*) As mechanical wave is produced when a source of energy creates a vibration that travels through a medium, such as air or water. A vibration is basically repeated back-and-forth motion, and it can be seen on the surface of a medium such as water. As the sound waves move through the water, a vibration visually disturbs the surface. The greater the strength of the sound wave, the more the water is disturbed. In today's experiment, you are going test if the distance from the energy source makes a different in amount of vibrations that are transferred through the water.
- ☞ **Results** – The students should see that the farther the music player is from the bowl, the less the water is disturbed by the vibrations.
- ☞ **Explanation** – As sound travels through the air, it exerts pressure on the surrounding medium. Over time, the pressure the waves exert is dampened and the amplitude, or loudness, of the sound wave is decreased. Eventually, the sound waves are not felt, or heard, anymore. In the experiment, the students saw this effect in action. As the music player was moved further away from the water bowl, the amplitude of the sound waves decreased, which caused the overall effect of the vibrations in the water to be decreased.
- ☞ **Take it Further** – Have the students try different types of music to see how the different tones and rhythms affect the amount of vibrations. (*The louder and faster the music, the students will see more evidence of vibrations in the water.*)

Discussion Questions

Measuring Sound, pg. 180

1. Does sound travel at different speeds? (*No, all sound travels at the same speed.*)
2. What determines the loudness of a sound? (*The loudness of a sound is determined by the amplitude of the sound wave. The larger the amplitude, the louder the sound.*)
3. What causes the pitch of a sound to change? (*The pitch of a sound is determined by the frequency of the sound waves. The closer together the sound waves are, the higher the frequency and the higher the pitch of the sound we hear.*)

Sound Recording, pg. 188

1. What are sound recordings? (*Sound recordings are simply copies of sound waves.*)
2. What is the difference between analog and digital recordings? (*Analog recordings store the sound waves on records or strips of tape as wavy lines. Digital recordings change the sound wave patterns into numbers, which show the position of the wave. Then, these numbers are stored as tiny pits on a CD or magnetic patterns.*)

Want More

- **Mechanical Wave Speed Worksheet** – Have the students practice using the wave equation to calculate the speed of a mechanical wave with the worksheet in the Appendix on pg. 258.
  Answers:
  1. 2.1 m/s
  2. 1 m/s

3. 56.72 Hz
4. 0.375 km (or 375 m)

- **Sound Simulation -** Have the students do the following on-line sound simulation from PhET:
  - http://phet.colorado.edu/en/simulation/sound
- **Doppler effect –** Have the students watch the following video about the Doppler Effect.
  - https://www.youtube.com/watch?v=h4OnBYrbCjY

Sketch Week 14

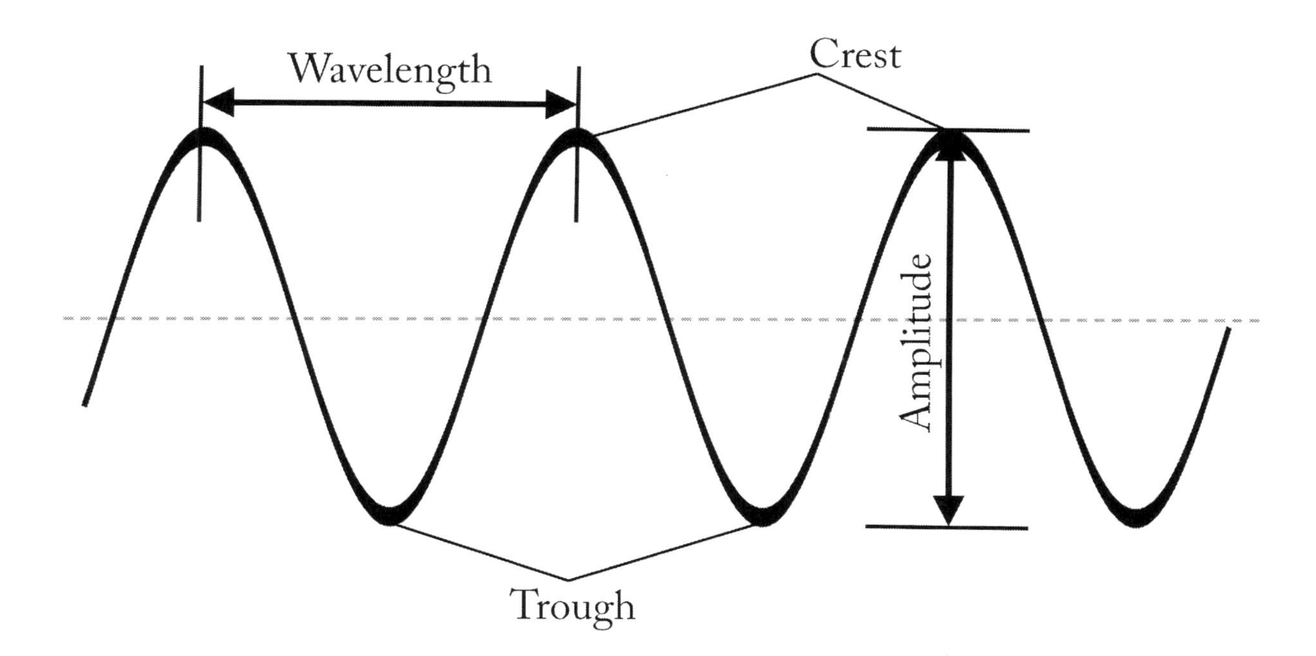

Student Assignment Sheet Week 15
Hearing Sound

Experiment: Make a Tonoscope

Materials

- ✓ Plastic jar, or small flower pot
- ✓ A piece of latex material large enough to cover the lid of your jar (like the kind used for exercise bands)
- ✓ 1" plastic tubing
- ✓ Rubber band
- ✓ Air-dry clay
- ✓ Salt

Procedure

1. Read the introduction to the experiment.
2. Cut a hole in the lower half of the plastic jar large enough to fit the tubing. Insert the tubing into the hole you just made. Then, pack a bit of clay around the edges so that air won't leak out.
3. Next, stretch the latex material over the opening of the jar and use the rubber band to secure it in place. Sprinkle a bit of salt over the rubber material.
4. Hum, sing, or speak through the tubing and watch what happens to the salt. Change the pitch of your voice and observe any changes. Write your observations on your experiment sheet.
5. Draw conclusions and complete the experiment sheet.

Vocabulary & Memory Work

- ☐ Vocabulary: antinoise, resonate frequency
- ☐ Memory Work—This week, continue working on memorizing the types of mechanical waves.

Sketch: Sound Production

- Draw a line representing the range of sounds the human, monkey, frog, and grasshopper can make. Label the lines with the following – 85 – 1,100 Hz (Human), 400 – 6,000 (Monkey), 50 – 8,000 Hz (Frog), and 7,000 – 100,000 Hz (Grasshopper).

Writing

- Reading Assignment: *DK Encyclopedia of Science* pg. 181 (Loudness), pp. 182-183 (Making and Hearing Sound)
- Additional Research Readings
  - 🕮 Electronic Sound: *DK EOS* pp. 189
  - 🕮 Resonance: *KSE* pp. 318-319

Dates

- 6th century BC – Pythagoras directly links the amplitude of the vibration of a plucked string to the perceived loudness of the instrument.

Schedules for Week 15

Two Days a Week

Day 1	Day 2
☐ Do the "Make a Tonoscope" experiment, and then fill out the experiment sheet on SG pp. 116-117 ☐ Define antinoise and resonate frequency on SG pg. 100 ☐ Enter the dates onto the date sheets on SG pp. 8-13	☐ Read pp. 181-183 from *DK EOS,* and then discuss what was read ☐ Color and label the "Sound Production" sketch on SG pg. 115 ☐ Prepare an outline or narrative summary; write it on SG pp. 118-119

Supplies I Need for the Week

✓ Plastic jar, or small flower pot
✓ A piece of latex material large enough to cover the lid of your jar
✓ 1" plastic tubing, Rubber band, Air-dry clay, Salt

Things I Need to Prepare

Five Days a Week

Day 1	Day 2	Day 3	Day 4	Day 5
☐ Do the "Make a Tonoscope" experiment, and then fill out the experiment sheet on SG pp. 116-117 ☐ Enter the dates onto the date sheets on SG pp. 8-13	☐ Read pp. 181-183 from *DK EOS,* and then discuss what was read ☐ Write an outline on SG pg. 118	☐ Define antinoise ande resonate frequency on SG pg. 100 ☐ Color and label the "Sound Production" sketch on SG pg. 115	☐ Read one or all of the additional reading assignments ☐ Write a report on what you learned on SG pg. 119	☐ Complete one of the Want More Activities listed **OR** ☐ Study a scientist from the field of Physics

Supplies I Need for the Week

✓ Plastic jar, or small flower pot
✓ A piece of latex material large enough to cover the lid of your jar (like the kind used for exercise bands)
✓ 1" plastic tubing, Rubber band, Air-dry clay, Salt

Things I Need to Prepare

Additional Information Week 15

Experiment Information

- **Introduction** – (*from the Student Guide*) A tonoscope is an acoustically device that allows you to see patterns in salt created by the sound of your voice. It was originally designed by Hans Jeny to explore wave behavior and patterns. In today's experiment, you are going to make your own tonoscope.
- **Results** – The students should see the salt crystals vibrate and move when they hum. As they change the pitch of their hum, they should that the salt crystals move to make different patterns, such as rings or waves.
- **Explanation** – The students' voice creates vibrations when they hum or talk into the tonoscope. As they change the pitch of their voices, they are able to tune into the different natural frequencies of the membrane causing it to resonate. Because of the variations in the latex, some parts of the membrane resonate while other don't. The parts that remain still are known as nodal points and the salt tends to move towards and collect at these points creating patterns on the tonoscope lid.
- **Troubleshooting** – Make sure that the rubber material is stretched tight. If the students find that the rubber band is not holding it securely in place, they can use tape or a cross stitch hoop instead.
- **Take it Further** – Have the students repeat the experiment, only this time have them place a few cotton balls in the bottom of the jar. (*The students should see that the cotton balls dampen the sound and decrease the effect of the vibrations on the salt.*)

Discussion Questions

Loudness , pg. 181

1. What does the loudness of a sound depend upon? (*The loudness of a sound depends upon the amount of energy the wave carries. The more energy the sound wave carries, the larger the amplitude and the louder the sound.*)
2. What does the decibel scale measure? (*The decibel scale measures the loudness of a sound.*)

Making and Hearing Sound, pp. 182-183

1. How do we hear sound? (*Our outer ears collect sound waves, which cause the eardrum to vibrate. The vibrations are then carried into the inner ear by a series of tiny bones. In the inner ear, fluid carries the vibrations through a narrow tube that stimulates the hairs lining the tube. This causes nerves to send electrical impulses to the brain, which we recognize as sound.*)
2. What is resonance? (*Every object has a point at which it naturally vibrates, called resonate frequency. Resonance is when a sound wave of the same frequency as the resonate frequency of an object causes the object to vibrate.*)
3. How does a voice activated device work? (*A voice activated device has a microphone that can detect sound above a certain level.*)

Want More

- **Resonance** – Have the students do the activity suggested on *DK Encyclopedia of Science* pg.

182, under the "Air in Bottles" section.

Sketch Week 15

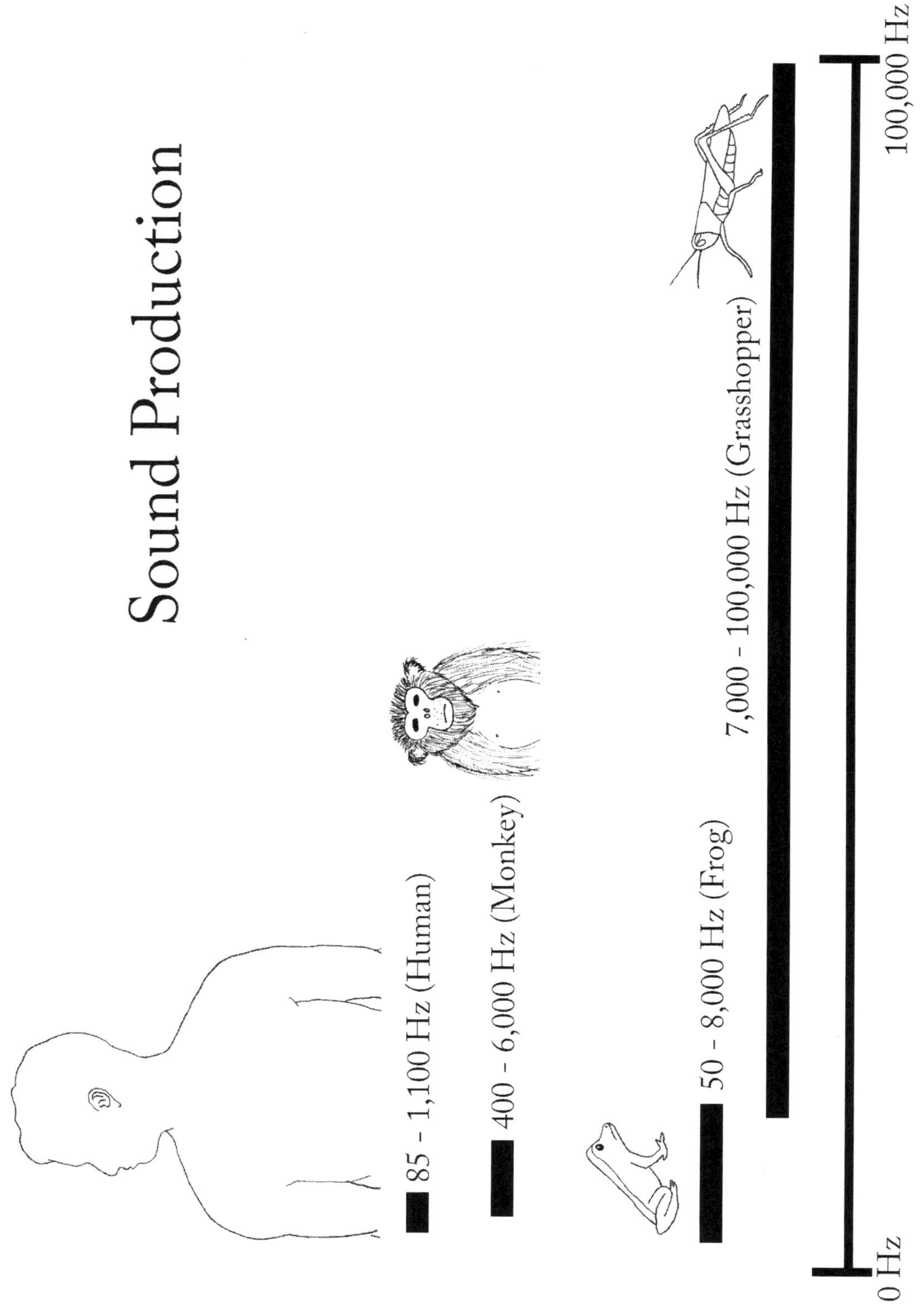

Student Assignment Sheet Week 16
Acoustics

Experiment: Does the size of the room I am in affect the way I hear?

Materials
- ✓ Partner
- ✓ Blindfold

Procedure
1. Read the introduction to the experiment and answer the question for the hypothesis section.
2. Find the largest room in your house, like the basement or garage. Stand in the center of the room and put on the blindfold. Have your partner stand in front, in back, or to the side of you. He or she should be about 3 feet (1 meter) from you with their back to you. Have your partner whisper your name and then try to determine his or her location. Repeat this process until your partner has stood in front, in back, and to both sides of you. Write down on your experiment sheet which locations you got correct.
3. Next find an average-sized room in your house, like a bedroom. Repeat the process from step 2.
4. Finally find the smallest room in your house, like a bathroom. Repeat the process from step 2.
5. Draw conclusions and complete the experiment sheet.

Vocabulary & Memory Work
- ☐ Vocabulary: acoustics, echolocation
- ☐ Memory Work—This week, continue working on memorizing the types of mechanical waves.

Sketch: Dolphin Echolocation
- Label the following – dolphin makes a clicking sound, sending energy into the surrounding water; soft surfaces absorb part of the sound energy and reflect the rest; hard surfaces reflect the sound energy; energy travels through the water

Writing
- Reading Assignment: *DK Encyclopedia of Science* pp. 184-185 (Reflection and Absorption)
- Additional Research Readings
  - Musical Sounds: *DK EOS* pp. 186-187
  - Vibrations of Strings: *KSE* pg. 320
  - Vibrations in Tubes: *KSE* pg. 321
  - Perception of Sound: *UIDS* pp. 42-43

Dates
- 1906 – American naval architect Lewis Nixon invents the first sonar-like listening device, which he used to detect icebergs.

Schedules for Week 16

Two Days a Week

Day 1	Day 2
☐ Do the "Does the size of the room I am in affect the way I hear?" experiment, and then fill out the experiment sheet on SG pp. 122-123 ☐ Define acoustics and echolocation on SG pg. 101 ☐ Enter the dates onto the date sheets on SG pp. 8-13	☐ Read pp. 184-185 from *DK EOS*, and then discuss what was read ☐ Color and label the "Dolphin Echolocation" sketch on SG pg. 121 ☐ Prepare an outline or narrative summary; write it on SG pp. 124-125

Supplies I Need for the Week
✓ Partner
✓ Blindfold

Things I Need to Prepare

Five Days a Week

Day 1	Day 2	Day 3	Day 4	Day 5
☐ Do the "Does the size of the room I am in affect the way I hear?" experiment, and then fill out the experiment sheet on SG pp. 122-123 ☐ Enter the dates onto the date sheets on SG pp. 8-13	☐ Read pp. 184-185 from *DK EOS*, and then discuss what was read ☐ Write an outline on SG pg. 124	☐ Define acoustics and echolocation on SG pg. 101 ☐ Color and label the "Dolphin Echolocation" sketch on SG pg. 121	☐ Read one or all of the additional reading assignments ☐ Write a report on what you learned on SG pg. 125	☐ Complete one of the Want More Activities listed **OR** ☐ Study a scientist from the field of Physics

Supplies I Need for the Week
✓ Partner
✓ Blindfold

Things I Need to Prepare

Additional Information Week 16

Experiment Information

- ☞ **Introduction –** (*from the Student Guide*) Acoustics is the study of sound and the way it travels in a given space. When designing a performance hall, architects will use certain materials in specific places to create the desired sound effects for the space. In today's experiment, you are going to test the acoustics of various rooms in and outside your home.
- ☞ **Results –** The students should see that the smaller the room, the easier it was to determine where their partner was located.
- ☞ **Explanation –** The acoustics of a room depend upon its size and the furniture or items in it. When sound waves are emitted from a source, they are absorbed by soft surfaces and reflected by hard ones. Typically, furniture absorbs some of the sound waves and the walls reflect it. As the sound waves travel, the pressure they exert decreases, lowering the volume of the sound we can hear. So the larger a room, the longer the sound waves have to travel before they are reflected. Thus, the sound waves emitted by the partner in the largest room were hardest to hear and use for pinpointing their location.
- ☞ **Take it Further –** Have the students compare the acoustics outside versus the acoustics inside. Do this by repeating the procedure from step 2 of the experiment outdoors. (*The students should see that it is much easier to hear indoors.*)

Discussion Questions

1. What happens when sound waves hit a hard surface? (*When sounds waves hit a hard surface, the wave are reflected and bounce back.*) A soft surface? (*When sound waves hit a soft surfaces, some or all of the waves are absorbed.*)
2. What are echoes? (*Echoes are sound reflections.*)
3. What does a parabolic dish do? (*A parabolic dish collects and concentrates sound.*)
4. How do dolphins and other animals use echolocation? (*Dolphins use echolocation to find fish and to locate underwater obstacles.*)
5. How does ultrasound and sonar work? (*Both ultrasound and sonar send ultrasound waves into a medium. Then, they collect the echoes, or reflections, and use this information to form a picture of what can be found in the medium.*)

Want More

- **Marco Polo –** Have the students play Marco Polo, a game that uses echolocation. In this game, one person is blindfolded. The blindfolded player cries out Marco and the other players respond with Polo. The blindfolded player then tries to tag another player who then becomes the next blindfolded Marco.
- **Field Trip –** Take the students to a music hall to watch a performance. While you are there, look around and observe the different structures within the room and their placement. Discuss how the different items affect the acoustics of the hall.

Sketch Week 16

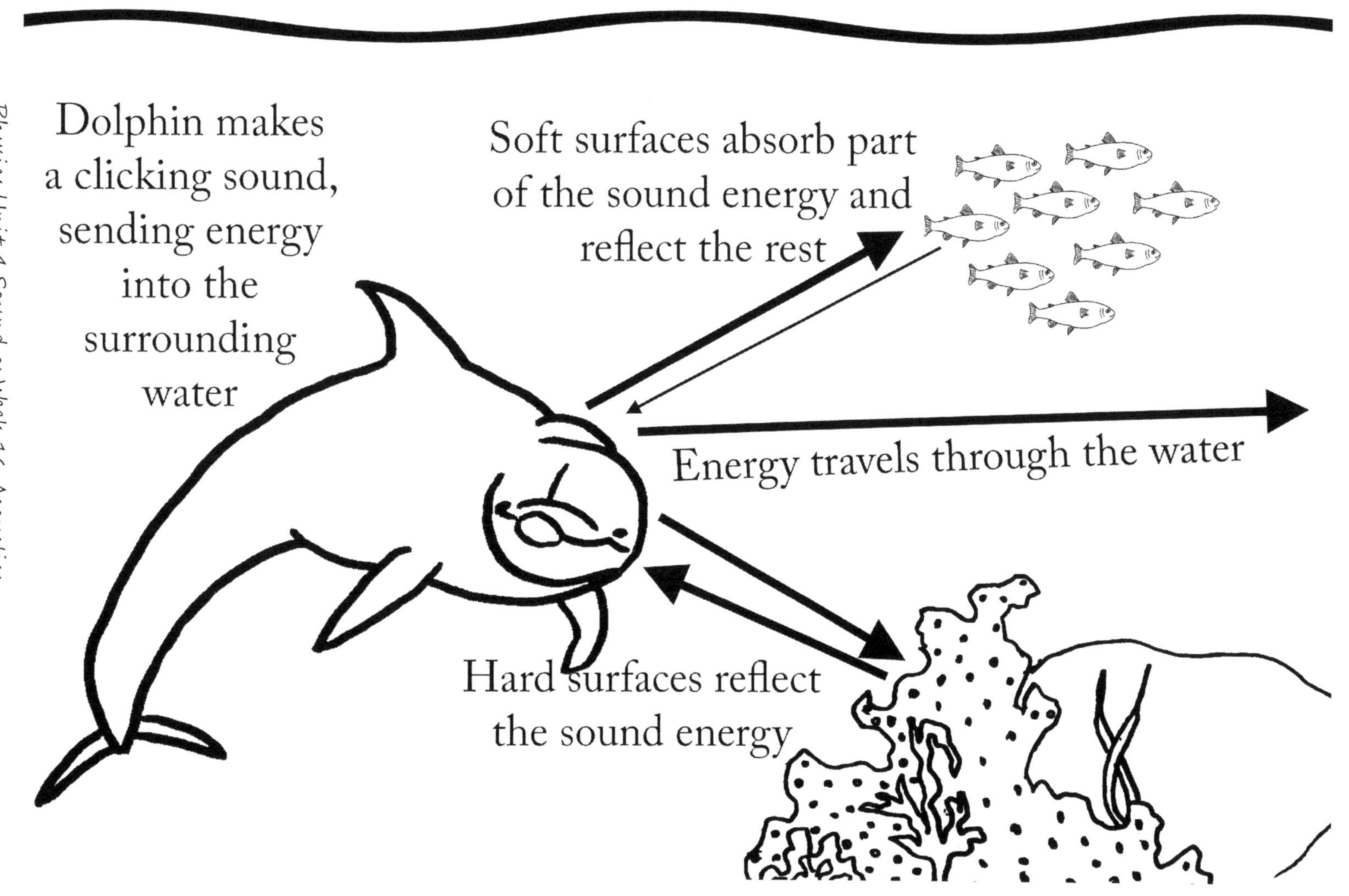

Unit 4 Sound
Unit Test Answers

Vocabulary Matching

1. B
2. F
3. A
4. J
5. H
6. E
7. C
8. G
9. I
10. D

True or False

1. True
2. False (*A sonic boom is caused when the sound of a supersonic jet catches up, causing a shock wave of sound.*)
3. False (*All sound waves travel at the same speed. The amplitude of the sound wave determines the loudness of the sound. The frequency of the waves determine the pitch of the sound.*)
4. True
5. False (*Resonance is when a sound wave of the same frequency as the resonate frequency of an object causes the object to vibrate.*)
6. True
7. True
8. False (*Dolphins use sound waves to locate fish and obstacles. Humans use ultrasound waves to map the bottom of the ocean.*)

Short Answer

1. Sound travels more quickly through a solid because the molecules are closer together and more firmly in place. So, they bounce back into place quicker than the molecules in a gas.
2. The loudness of a sound is determined by the amplitude of the sound wave. The larger the amplitude, the louder the sound.
3. Our outer ears collect sound waves, which cause the eardrum to vibrate. The vibrations are then carried into the inner ear by a series of tiny bones. In the inner ear, fluid carries the vibrations through a narrow tube that stimulates the hairs lining the tube. This causes nerves to send electrical impulses to the brain, which we recognize as sound.
4. When sounds waves hit a hard surface, the wave are reflected and bounce back. When sound waves hit a soft surfaces, some or all of the waves are absorbed.
5. The three main types of mechanical waves are:
   - ✓ **Transverse Wave –** A wave that causes the medium to vibrate perpendicular to the direction in which the wave travels.
   - ✓ **Longitudinal Waves –** A wave that causes the medium to vibrate parallel to the direction in which the wave travels.
   - ✓ **Surface Waves –** A wave that travels along the surface separating two media.

Unit 4 Sound
Unit Test

Vocabulary Matching

1. Mechanical wave ____
2. Sound ____
3. Vibration ____
4. Amplitude ____
5. Frequency ____
6. Wavelength ____
7. Antinoise ____
8. Resonate frequency ____
9. Acoustics ____
10. Echolocation ____

A. A quick back and forth movement, for example when sound waves causes a nearby glass of water to vibrate.

B. A wave that travels through a medium, such as air, water, or solids.

C. Produced when two sound waves overlap and cancel each other out.

D. The locating of objects through using reflected sound.

E. The distance between the crest of one wave to the crest of another.

F. A mechanical wave, or vibration, that travels through a medium, such as air, and can be heard when it reaches a person's or animal's ear.

G. The frequency at which an object naturally begins to vibrate.

H. The number a waves that pass a given point in a second.

I. The study of how sound travels in a given space.

J. The size of a vibration or the height of a wave.

True or False

1. ____________________ Sound is caused by the rapid motion of particles of matter colliding with one another, known as vibration.

2. ____________________ A sonic boom is cause when the sound of a supersonic jet precedes its appearance, causing a shock wave of sound.

3. ____________________ The loudness and pitch of a sound is determined by the speed at which the sound wave travels.

4. ______________________ Sound recordings are copies of sound waves.

5. ______________________ Resonance is when a sound wave of the different frequency as the resonate frequency of an object causes the object to vibrate.

6. ______________________ The loudness of a sound depends upon the amount of energy the wave carries.

7. ______________________ Echoes are sound reflections.

8. ______________________ Sounds waves are not useful in locating objects underwater.

Short Answer

1. Why does sound travel more quickly through a solid than a gas?

2. What determines the loudness of a sound?

3. How do we hear sound?

4. What happens when sound waves hit a hard surface? A soft surface?

5. What are the three main types of mechanical waves?

Physics Unit 5

Light

Unit 5 Light
Overview of Study

Sequence of Study

Week 17: Light
Week 18: Reflection and Refraction
Week 19: Vision and Color
Week 20: Optics

Materials by Week

Week	Materials
17	9 Ultraviolet light detecting beads, 3 Shallow dishes (not clear plastic or glass), Plastic Wrap, Two different levels of SPF sunscreen (i.e., SPF 15 and SPF 45), Rubber bands
18	4 Pencils, 4 Clear glasses, Water, Oil, Alcohol, Corn syrup
19	Thin cardboard, Red, blue, and yellow paint, 6 Rubber bands, Hole punch
20	Jell-O™ (orange, lemon, or lime), Round bowl or jar – at least 4" (10 cm) in diameter, 1 Cup water, Dull knife, Plate, Flashlight

Vocabulary for the Unit

1. **Electromagnetic waves** – These waves carry energy from one place to another and are produced when an electric charge vibrates or accelerates.
2. **Electromagnetic spectrum** – The full range of electromagnetic radiation, which includes radio waves, infrared rays, visible light, Ultraviolet light, X-rays, and Gamma rays.
3. **Photons** – Packets of electromagnetic energy.
4. **Reflection** – The bouncing back of light from a surface.
5. **Refraction** – The change in direction of a light beam as it passes from one medium to another of different density.
6. **Convex** – A lens that curves outward.
7. **Concave** – A lens that curves inward.
8. **Diffraction** – The spreading out of light waves when they pass through a narrow slit.
9. **Interference** – The disturbance of a signal when two or more waves meet.
10. **Converging lens** – A lens that causes parallel light rays passing through it to come together.
11. **Diverging lens** – A lens that causes parallel light rays passing through it to spread out.
12. **Lens** – A piece of transparent material with a curved edge that causes light to bend in a particular way.

Memory Work for the Unit

Waves of the Electromagnetic Spectrum

1. **Radio waves** – These waves are typically found in radios, televisions, microwaves, and radars.
2. **Infrared rays** – These waves are used as a source of heat as well as to find areas of heat differences.
3. **Visible light** – These waves make up the spectrum of visible light we can see.
4. **Ultraviolet rays** – These waves can kill microorganisms, produce vitamin D in our skin, and help plants grow.
5. **X-rays** – These waves have very short wavelengths and are used to make pictures of what is inside a solid object.
6. **Gamma rays** – These waves have the shortest wavelengths in the electromagnetic spectrum. Gamma rays are used in medicine to kill cancer cells and to provide images of the brain.

Equation

- Speed of Waves Equation

  $v = \lambda \bullet f$

  "v" stands for speed.
  "λ" stands for wavelength.
  "f" stands for frequency.

Notes

Student Assignment Sheet Week 17
Light

Experiment: Does sunscreen really block UV rays?

Materials

- ✓ 9 Ultraviolet light detecting beads
- ✓ 3 Shallow dishes (not clear plastic or glass)
- ✓ Plastic Wrap
- ✓ Two different levels of SPF sunscreen (i.e., SPF 15 and SPF 45)
- ✓ Rubber bands

Procedure

1. Read the introduction to the experiment and answer the question for the hypothesis section.
2. Place three of the ultraviolet detecting beads in each of the bowls and label them a "A," "B," and "C." Then, cut the plastic wrap to cover bowl each one and secure it in place with rubber band or tape.
3. Next, slather a fair amount of the lower SPF sunscreens over the plastic wrap paper on bowl "B," and a fair amount of the higher SPF sunscreen over the plastic wrap on bowl "C." Place on three bowls on a tray, take them outside, and set them on a table in the full sun.
4. Check the beads after 30 seconds, 2 minutes, 5 minutes, 10 minutes, and 30 minutes. Each time, rate the degree of color change on a scale of 1 to 10 (e.g. If you beads have changed color by about 20%, rate them a 2). Record your observations each time on your experiment sheet.
5. Draw conclusions and complete the experiment sheet.

Vocabulary & Memory Work

- ☐ Vocabulary: electromagnetic waves, electromagnetic spectrum, photons
- ☐ Memory Work—This week, begin working on memorizing the waves of the electromagnetic spectrum and the speed of waves equation (if you did not memorize it in the last unit).

Sketch: Light

- Label the following – source of light; at times, light behaves as if it is composed of a stream of particles; at times, light behaves as if it is a wave motion.

Writing

- Reading Assignment: *DK Encyclopedia of Science* pp. 190-191 (Light), pg. 192 (Electromagnetic Spectrum)
- Additional Research Readings
  - Sources of Light: *DK EOS* pg. 193
  - Light and Matter: *DK EOS* pg. 200
  - Light: *KSE* pp. 260-261, *UIDS* pg. 46

Dates

- 1672 – Isaac Newton suggests that light is composed of tiny particles resembling balls.
- 1678 – Christian Huygens suggests that light is a wave motion, similar to sound or water waves.
- 1900 – Max Planck suggests that light is a combination of a particle and a wave, forming the basis of quantum theory.
- 1905 – Einstein expands upon Planck's work and suggests that light is composed of tiny particles, called photons, which have energy and behave like waves.

Schedules for Week 17

Two Days a Week

Day 1	Day 2
☐ Do the "Does sunscreen really block UV rays?" experiment, and then fill out the experiment sheet on SG pp. 132-133 ☐ Define electromagnetic waves, electromagnetic spectrum, and photons on SG pg. 128 ☐ Enter the dates onto the date sheets on SG pp. 8-13	☐ Read pp. 190-192 from *DK EOS,* and then discuss what was read ☐ Color and label the "Light" sketch on SG pg. 131 ☐ Prepare an outline or narrative summary; write it on SG pp. 134-135
Supplies I Need for the Week ✓ 9 Ultraviolet light detecting beads ✓ 3 Shallow dishes (not clear plastic or glass) ✓ Two different levels of SPF sunscreen (i.e., SPF 15 and SPF 45) ✓ Plastic Wrap, Rubber bands	
Things I Need to Prepare	

Five Days a Week

Day 1	Day 2	Day 3	Day 4	Day 5
☐ Do the "Does sunscreen really block UV rays?" experiment, and then fill out the experiment sheet on SG pp. 132-133 ☐ Enter the dates onto the date sheets on SG pp. 8-13	☐ Read pp. 190-192 from *DK EOS,* and then discuss what was read ☐ Write an outline on SG pg. 134	☐ Define electromagnetic waves, electromagnetic spectrum, and photons on SG pg. 128 ☐ Color and label the "Light" sketch on SG pg. 131	☐ Read one or all of the additional reading assignments ☐ Write a report on what you learned on SG pg. 135	☐ Complete one of the Want More Activities listed **OR** ☐ Study a scientist from the field of Physics
Supplies I Need for the Week ✓ 9 Ultraviolet light detecting beads ✓ 3 Shallow dishes (not clear plastic or glass) ✓ Two different levels of SPF sunscreen (i.e., SPF 15 and SPF 45) ✓ Plastic Wrap, Rubber bands				
Things I Need to Prepare				

Additional Information Week 17

Experiment Information

- **Introduction** – (*from the Student Guide*) Ultraviolet light from the Sun is beneficial. These electromagnetic waves are necessary for our skin to produce Vitamin D, and the waves help plants to grow. However, prolonged exposure to ultraviolet light rays can cause sunburn, wrinkles, and eventually skin cancer. In today's experiment, you are going to test to see whether or not sunscreen helps to block Ultraviolet light rays.
- **Results** – The students should see that the beads in the bowl "A" changed color almost immediately. The beads in bowl "B" and "C" should have started to change color a bit within the first few minutes, but may have not fully changed color by the end of the thirty minutes. The students should have also seen that the color change in the bowl "B" beads was more dramatic than the color change in bowl "C."
- **Explanation** – The UV beads contain pigments that change color when exposed to UV rays in the sunlight. The UV light rays are able to penetrate the plastic wrap with ease, so the beads in bowl "A" change color fully within seconds. The sunscreens on the plastic wraps contains SPF which blocks UV rays. The higher the SPF, the more efficient the sunscreen is at blocking the rays and the longer the effect will last. This is why the students saw that the beads in bowl "C" took the longest to change color fully.
- **Troubleshooting** – If the students' beads are changing color too fast:
  - Make sure that there is an even coating of sunscreen on the plastic wrap.
  - Make sure that the container is not see-through.

  If the students' beads are not changing color at all:
  - Make sure that they are in direct sunlight.
- **Take it Further** – Have the students repeat the experiment by comparing spray-on and lotion sunscreens of the same SPF levels to see if the sunscreen type makes a difference. They can also test if old sunscreen blocks as well as new sunscreen. (*They should see that an even coat of spray-on sunscreen works just as well as the lotion. They should also see that old sunscreen does not work as well as new screen because the SPF chemicals begin to break down.*)

Discussion Questions

Light, pp. 190-191

1. What is light? (*Light is a form of energy.*)
2. How does light differ from sound? (*Light can travel in a vacuum, but sound cannot.*)
3. What is the photoelectric effect? (*The photoelectric effect says that when light shines on a metal, it has the ability to knock electrons out of the metal. This principle is the basis of solar power.*)
4. What does quantum theory say about light? (*Quantum theory suggests that light has a combination of the properties of a wave and of a particle.*)

Electromagnetic Spectrum, pg. 192

1. How are electromagnetic waves similar? (*Electromagnetic waves all travel at the speed of light.*) How do they differ? (*Each group of electromagnetic waves has different wavelengths and carries different amounts of energy.*)

2. Which electromagnetic waves have shorter wavelengths and carry more energy that visible light? (*Ultraviolet, X-ray, and gamma rays have shorter wavelengths and carry more energy that visible light.*) Which ones have longer wavelengths and carry less energy than visible light? (*Infrared, microwave, and radio waves have longer wavelengths and carry less energy than visible light.*)

Want More

- **Young's Experiment** – Have the students watch the following video from Khan Academy on Young's double-slit experiment, which shows how light behaves like a wave.
  - https://www.khanacademy.org/partner-content/mit-k12/mit-k12-physics/v/thomas-young-s-double-slit-experiment
- **Electromagnetic Wavelength Worksheet** – Have the students practice using the wave equation to calculate the speed of electromagnetic waves with the worksheet in the Appendix on pg. 259.
  Answers:
  1. 1.0×10^8 Hz
  2. 5×10^7 Hz
  3. 15.8 m

Sketch Week 17

Light

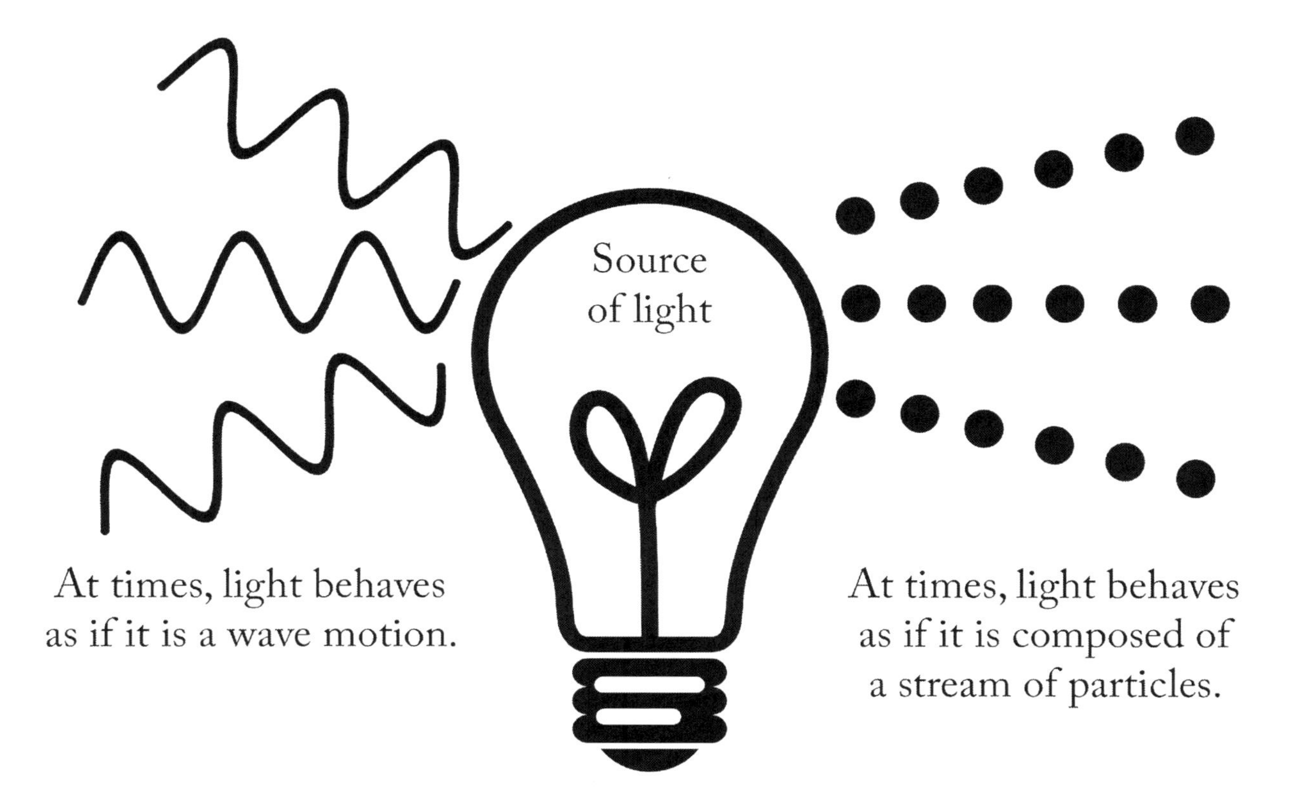

Student Assignment Sheet Week 18
Reflection and Refraction

Experiment: Which liquid splits a pencil the most?

Materials
- ✓ 4 Pencils
- ✓ 4 Clear glasses
- ✓ Water
- ✓ Oil
- ✓ Alcohol
- ✓ Corn syrup

Procedure
1. Read the introduction to the experiment and answer the question for the hypothesis section.
2. Label the cups #1 through #4 and place a pencil in each cup. Observe how the pencil looks in the cup without any liquid.
3. Now add ½ cup (about 120 mL) of water to cup #1, add ½ cup (about 120 mL) of oil to cup #2, ½ cup (about 120 mL) of alcohol to cup #3, and ½ cup (about 120 mL) of corn syrup to cup #4.
4. Observe what happens to the pencils in the glasses and write your observations down on your experiment sheet.
5. Draw conclusions and complete the experiment sheet.

Vocabulary & Memory Work
- ☐ Vocabulary: reflection, refraction
- ☐ Memory Work—This week, continue to work on memorizing the waves of the electromagnetic spectrum.

Sketch: Reflection
- Label the following – Reflection, mirror, angle of incidence, angle of reflection, the law of reflection states that the angle of reflection is equal to the angle of incidence
- Draw arrows to show what happens to the light rays when they are reflected and when they are refracted.

Writing
- Reading Assignment: *DK Encyclopedia of Science* pp. 194-195 (Reflection), pg. 196 (Refraction)
- Additional Research Readings
  - Reflection: *KSE* pp. 262-263, *UIDS* pp. 47-49
  - Refraction: *KSE* pp. 264-265, *UIDS* pp. 50-53

Dates
- 1902 – Hendrick Lorentz is awarded the Nobel Peace Prize for his work with electromagnetic waves and the propagation of light.

Schedules for Week 18

Two Days a Week

Day 1	Day 2
☐ Do the "Which liquid splits a pencil the most?" experiment, and then fill out the experiment sheet on SG pp. 138-139 ☐ Define reflection and refraction on SG pg.128 ☐ Enter the dates onto the date sheets on SG pp. 8-13	☐ Read pp. 194-196 from *DK EOS,* and then discuss what was read ☐ Color and label the "Reflection" sketch on SG pg. 137 ☐ Prepare an outline or narrative summary; write it on SG pp. 140-141
Supplies I Need for the Week ✓ 4 Pencils, 4 Clear glasses ✓ Water, Oil, Alcohol, Corn syrup	
Things I Need to Prepare	

Five Days a Week

Day 1	Day 2	Day 3	Day 4	Day 5
☐ Do the "Which liquid splits a pencil the most?" experiment, and then fill out the experiment sheet on SG pp. 138-139 ☐ Enter the dates onto the date sheets on SG pp. 8-13	☐ Read pp. 194-196 from *DK EOS,* and then discuss what was read ☐ Write an outline on SG pg. 140	☐ Define reflection and refraction on SG pg. 128 ☐ Color and label the "Reflection" sketch on SG pg. 137	☐ Read one or all of the additional reading assignments ☐ Write a report on what you learned on SG pg. 141	☐ Complete one of the Want More Activities listed **OR** ☐ Study a scientist from the field of Physics
Supplies I Need for the Week ✓ 4 Pencils, 4 Clear glasses ✓ Water, Oil, Alcohol, Corn syrup				
Things I Need to Prepare				

Additional Information Week 18

Experiment Information

- **Introduction –** (*from the Student Guide*) Light typically travels in a straight line. However, as it passes through different media, it bends slightly. This phenomenon is known as refraction and it occurs because light travels at different speeds through different media. In today's experiment, you are going to test how different liquids refract the reflected light from a pencil is refracted in different liquids.
- **Results –** The pencil should not appear to be split in the cup by itself. The pencils should appear to be split by all four of the liquids. The degree of split will be greater in the oil and corn syrup.
- **Explanation –** The refractive index of a material is a number that measures how much a light beam bends in a given material. The refractive index of water is 1.3, alcohol is 1.35, oil is 1.5, and corn syrup is 1.4. Each of the materials materials bends the reflected light of the pencil, causing the image to distort and look like the pencil was split. The oil and the corn syrup appeared to split the pencil farther because their reflective indexes are higher than those of water and alcohol.
- **Take it Further –** Have the students use a ruler to draw a line on an index card. Then, have them place a half-full glass of water on top of the card. (*The students should see that the line looks like it has been bent.*) You can have the students repeat the process with alcohol, corn syrup, and oil.

Discussion Questions

Reflection, pp. 194-195

1. How do we see objects that don't make their own light? (*Objects that don't make light reflect it so that we can see them.*)
2. What happens when a surface does not reflect light? (*When a surface reflects light, it looks black.*)
3. What causes a mirror image? (*A mirror image appears when the reflected light travels to our eyes as if it had come from behind the mirror.*)
4. What kind of images do convex mirrors produce? (*Convex mirrors always produce an image that is smaller.*) Concave mirrors? (*Concave mirrors always produce an image that is magnified.*)
5. What is the law of reflection? (*The law of reflection states that the angle of reflection is equal to the angle of incidence.*)

Refraction, pg. 196

1. What happens when light travels from one transparent material to another? (*When light travels from one transparent material to another, it bends. This is known as refraction.*)
2. What is the refractive index of a material? (*The refractive index of a material is a number that give the relationship between the speeds of light in two different media.*)

Want More

- **Bending Light –** Have the students do the following online tutorial from PhET:
  - https://phet.colorado.edu/en/simulation/bending-light

Mirrors – Have the students observe reflection. You will need a mirror, flashlight, and dark room. Have the students go into the dark room, shine the flashlight at the mirror, and observe where the beam of light ends up. (*The students should see that the beam of light is reflected onto the opposite wall.*)

Sketch Week 18

Reflection

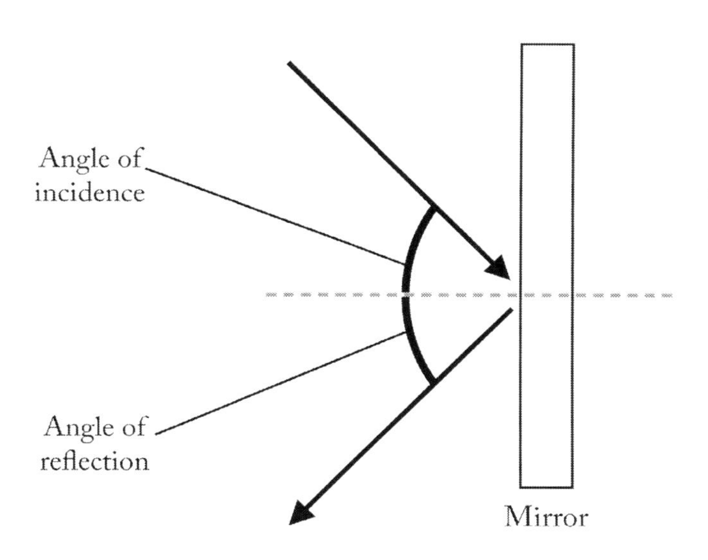

The law of reflection states that the angle of reflection is equal to the angle of incidence.

Student Assignment Sheet Week 19
Vision and Color

Experiment: Can I trick my brain and change the color?

Materials

- ✓ Thin cardboard
- ✓ Red, blue, and yellow paint
- ✓ 6 Rubber bands
- ✓ Hole punch

Procedure

1. Read the introduction to the experiment and answer the questions in the hypothesis section.
2. Cut out 6 circles of the same size from the cardboard. Paint two of them red, two of them yellow, and two of them blue. Set them aside to dry.
3. Once they are dry, glue one of the red disks to one of the yellow disks, one of the red disks to one of the blue disks, and one of the blue disks to one of the yellow disks. Then, punch holes on opposite sides of the three disks and tie a rubber band into each hole.
4. Take the red/blue disk, twist the rubber bands, and then stretch them out so that the disk spins quickly. Observe what color you see and write it on your experiment sheet.
5. Repeat step 4 with the red/yellow and blue/yellow disks.
6. Draw conclusions and complete the experiment sheet.

Vocabulary & Memory Work

- ☐ Vocabulary: convex, concave, diffraction, interference
- ☐ Memory Work—This week, continue to work on memorizing the waves of the electromagnetic spectrum.

Sketch: Spectrum of Visible Light

- Label the following – Visible light; Prism splits visible light into its spectral components; Red light, Wavelength 610 to 750 nm; Orange light, Wavelength 590 to 610 nm; Yellow light, Wavelength 570 to 590 nm; Green light, Wavelength 500 to 570 nm; Blue light, Wavelength 450 to 500 nm; Violet light, Wavelength 400 to 450 nm.

Writing

- Reading Assignment: *DK Encyclopedia of Science* pg. 202 (Color), pp. 204-205 (Vision)
- Additional Research Readings
  - 🕮 Color Subtraction: *DK EOS* pg. 203
  - 🕮 Color: *KSE* pp. 272-273

Dates

- There are no dates to be entered this week.

Schedules for Week 19

Two Days a Week

Day 1	Day 2
☐ Do the "Can I trick my brain and change the color?" experiment, and then fill out the experiment sheet on SG pp. 144-145 ☐ Define convex, concave, diffraction, and interference on SG pg. 128-129 ☐ Enter the dates onto the date sheets on SG pp. 8-13	☐ Read pp. 202, 204-205 from *DK EOS,* and then discuss what was read ☐ Color and label the "Spectrum of Visible Light" sketch on SG pg. 143 ☐ Prepare an outline or narrative summary; write it on SG pp. 146-147
Supplies I Need for the Week ✓ Thin cardboard ✓ Red, blue, and yellow paint ✓ 6 Rubber bands, Hole punch	
Things I Need to Prepare	

Five Days a Week

Day 1	Day 2	Day 3	Day 4	Day 5
☐ Do the "Can I trick my brain and change the color?" experiment, and then fill out the experiment sheet on SG pp. 144-145 ☐ Enter the dates onto the date sheets on SG pp. 8-13	☐ Read pp. 202, 204-205 from *DK EOS,* and then discuss what was read ☐ Write an outline on SG pg. 146	☐ Define convex, concave, diffraction, and interference on SG pg. 128-129 ☐ Color and label the "Spectrum of Visible Light" sketch on SG pg. 143	☐ Read one or all of the additional reading assignments ☐ Write a report on what you learned on SG pg. 147	☐ Complete one of the Want More Activities listed **OR** ☐ Study a scientist from the field of Physics
Supplies I Need for the Week ✓ Thin cardboard ✓ Red, blue, and yellow paint ✓ 6 Rubber bands, Hole punch				
Things I Need to Prepare				

Additional Information Week 19

Experiment Information

- ☞ **Introduction –** (*from the Student Guide*) Visible light is made up of a spectrum of wavelengths. When we see color, we are actually see the wavelength(s) of light that are reflected by the object. Light receptors in our eyes receive the reflected wavelengths and then our brain translated into the sensation of color. In today's experiment, you are going to try to trick your brain into seeing different colors than are actually there.
- ☞ **Results –** When the students spin the red/blue disk, they will see purple. When the students spin the red/yellow disk, they will see orange. When the students spin the blue/yellow disk, they will see green.
- ☞ **Explanation –** We know that when we mix the primary colors of red and blue we make purple. Obviously, the colors on the two opposite disks are not physically mixing. Instead, the disks are moving so fast that the students' brains are unable to process the individual colors of the disks. The brain creates a shortcut and combines the two wavelengths together to create the perception of seeing the secondary color.
- ☞ **Troubleshooting** – If the students can still see both colors, the disks are not spinning fast enough.
- ☞ **Take it Further –** Have the students make a disk with alternating sections of red, blue, and green. Have them poke a hole in the center and put a sharpened pencil in the hole. Then, have the students spin the disk like a top and observe what happens. (*The students should see that the colors blend together. If the disk spins fast enough, it will appear white.*)

Discussion Questions

Color, pg. 202

1. What makes a color? (*A color is formed by a wavelength or combination of wavelengths of light.*)
2. What happens when all the wavelengths of visible light are mixed? (*When all the wavelengths of visible light are mixed, the result is white light.*)
3. What happens in a prism? (*In a prism, a beam of light is split and spread out. Red light is refracted the least, while violet light is refracted to the greatest degree.*)

Vision, pp. 204-205

1. How do the eyes and brain work together to allow us to see? (*The lens of the eye focuses the light on the retina, which has light-sensitive cells. These cells send signals through the optic nerve to the brain. The brain interprets the signals and builds a picture that we can see.*)
2. How can a concave lens help to correct vision? (*A concave lens helps to correct far-sightedness by concentrating light rays.*) A convex one? (*A convex lens helps to correct near-sightedness by spreading out light rays.*)
3. Why do we see colors differently during the day and night? (*Our eyes have rods, which detect low light levels, and cones, which detect color. During the day, both the rods and cones are activated by the light, so we can see colors better. During the night, only the rods are stimulated, so colors are not as distinct.*)

Want More

- **Color Vision** – Have the students do the following online tutorial from PhET:
  - https://phet.colorado.edu/en/simulation/color-vision

Sketch Week 19

The sketch information is not found in the spine this week, but it is important for the students to know. It should be relatively easy for them to figure out, given what they know, but be aware that they may need your assistance this week.

Spectrum of Visible Light

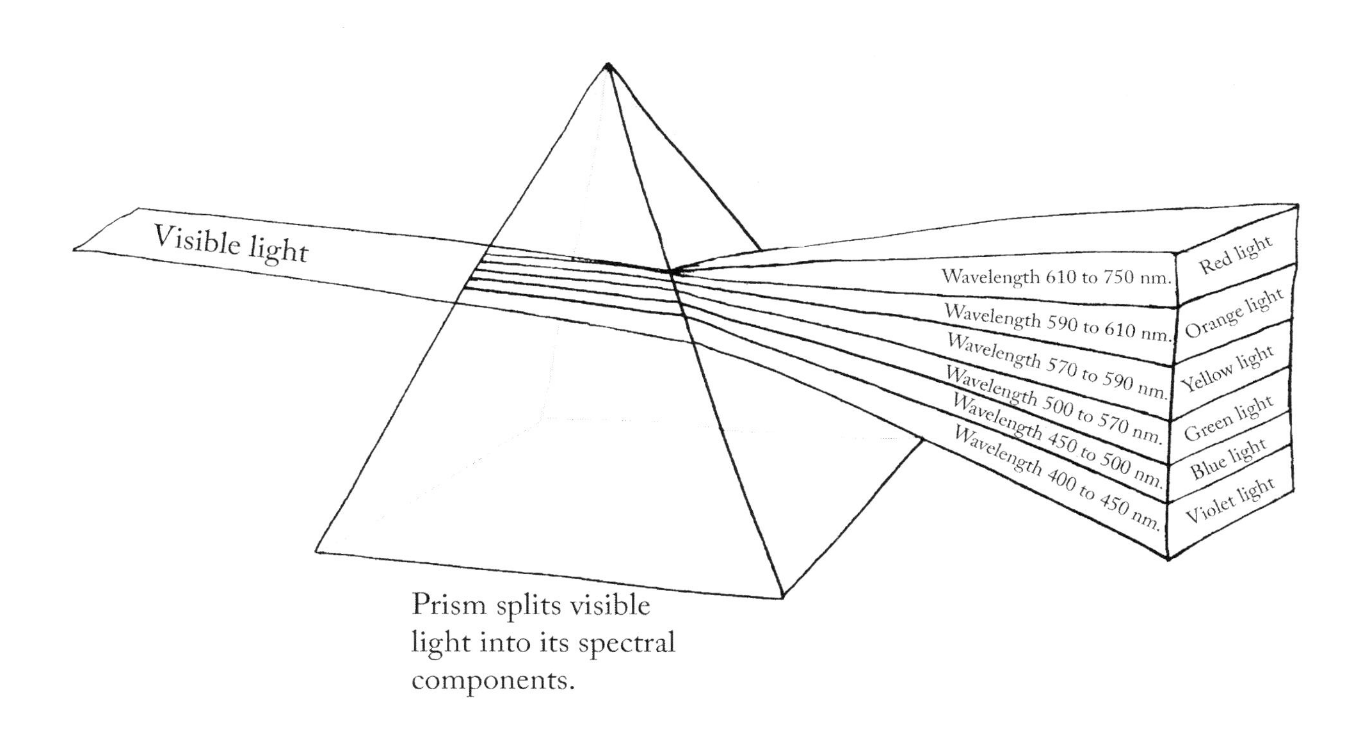

Prism splits visible light into its spectral components.

Student Assignment Sheet Week 20
Optics

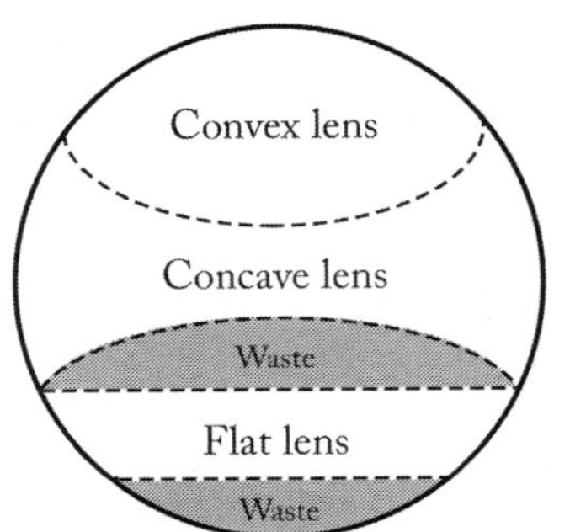

Experiment: How does shape affect the way a lens projects light?

Materials

- ✓ 1 Package Jell-O™ (orange, lemon, or lime)
- ✓ Round, flat-bottomed bowl or jar – at least 4" (10 cm) in diameter
- ✓ 1 Cup water, Dull knife, Plate, Flashlight

Procedure

****Note**—The night before you do this experiment, mix the package of Jell-O with 1 cup (240 mL) of very hot water. Pour a 1½ inch (about 4 cm) layer of the Jell-O into the round, flat-bottomed jar sprayed with a bit of oil. Set it in the fridge over night to set.**

1. Read the introduction to the experiment and answer the question for the hypothesis section.
2. Take the jar out and set it in warm water for 30 seconds to a minute. Gently remove the Jell-O disk and set it on a plate. Cut it into three lenses (concave, flat, and convex) using the diagram above.
3. Move to an interior room with no windows and place the plate about a foot away from the wall.
4. Set the flat lens on the edge of the plate, place the small flashlight behind it, turn the flashlight on, and then turn out the lights. Observe the light pattern that is displayed and measure from one end of the display to the other. Turn the light back on and record this information on your experiment sheet.
5. Next, set the concave lens on the edge of the plate so that the curved part faces the flashlight, turn the flashlight on, and then turn out the lights. Observe the light pattern that is displayed and measure from one end of the display to the other. Turn the light back on and record this information on your experiment sheet.
6. Repeat step 6 with the convex lens. Draw conclusions and complete the experiment sheet.

Vocabulary & Memory Work

- ☐ Vocabulary: converging lens, diverging lens, lens

Sketch: Convex vs. Concave

- Label the following – concave lens, convex lens, light.
- Draw the rays of light as they enter and exit the concave and convex lenses.

Writing

- Reading Assignment: *DK Encyclopedia of Science* pg. 197 (Lenses), pg. 198 (Optical Instruments)
- Additional Research Readings
  - Lenses and Curved Mirrors: *KSE* pp. 266-267
  - Optical Instruments: *UIDS* pp. 54-55

Dates

- Late 1600's – Antoni Van Leeuwenhoek designs the first microscope.
- 1789 – Will Herschel designs a telescope with a four foot diameter.
- 1992 – The Keck telescope is built, which is thirty-three feet in diameter.

Schedules for Week 20

Two Days a Week

Day 1	Day 2
☐ Do the "How does shape affect the way a lens projects light?" experiment, and then fill out the experiment sheet on SG pp. 150-151 ☐ Define converging lens, diverging lens, and lens on SG pg. 129 ☐ Enter the dates onto the date sheets on SG pp. 8-13	☐ Read pp. 197-198 from *DK EOS,* and then discuss what was read ☐ Color and label the "Convex vs. Concave" sketch on SG pg. 149 ☐ Prepare an outline or narrative summary; write it on SG pp. 152-153

Supplies I Need for the Week

- ✓ 1 Package Jell-O (orange, lemon, or lime)
- ✓ Round, flat-bottomed bowl or jar – at least 4" (10 cm) in diameter
- ✓ 1 Cup water, Dull knife, Plate, Flashlight

Things I Need to Prepare

Five Days a Week

Day 1	Day 2	Day 3	Day 4	Day 5
☐ Do the "How does shape affect the way a lens projects light?" experiment, and then fill out the experiment sheet on SG pp. 150-151 ☐ Enter the dates onto the date sheets on SG pp. 8-13	☐ Read pp. ~~1897-198~~ 197 198 from *DK EOS,* and then discuss what was read ☐ Write an outline on SG pg. 152	☐ Define converging lens, diverging lens, and lens on SG pg. 129 ☐ Color and label the "Convex vs. Concave" sketch on SG pg. 149	☐ Read one or all of the additional reading assignments ☐ Write a report on what you learned on SG pg. 153	☐ Complete one of the Want More Activities listed **OR** ☐ Study a scientist from the field of Physics

Supplies I Need for the Week

- ✓ 1 Package Jell-O (orange, lemon, or lime)
- ✓ Round, flat-bottomed bowl or jar – at least 4" (10 cm) in diameter
- ✓ 1 Cup water, Dull knife, Plate, Flashlight

Things I Need to Prepare

Additional Information Week 20

Experiment Information

- **Introduction –** (*from the Student Guide*) A lens is a piece of transparent material that causes light to bend in a particular way. Lenses are used in glasses, photography, telescopes, and more. They are used to produce images, to magnify objects, to focus light, and to reduce images in a scene. In today's experiment, you are going to examine how the shape of a lens affects how it projects light.
- **Results –** The students should see that the flat lens create a pattern of light on the wall. The concave lens caused the pattern of light to come closer together. The convex lens caused the pattern of light to spread out.
- **Explanation –** The flat lens allows light to pass though in about the same way it entered, so the light rays are not scattered in any direction. The concave lens, which curves inward, caused the light rays to be scattered outward. This divergence of the light rays caused the image that was produced to be spread out and magnified. The convex lens, which curves outward, cause the light rays to be scattered inward. This convergence of the light rays caused the image that was produced to be compacted.
- **Troubleshooting –** Here are a few tips for working with the Jell-O lens:
  - Be sure to coat the jar or bowl with oil so that the Jell-O can be easily removed.
  - If you find the Jell-O is splitting when you cut it, warm up the knife a bit by running it under hot water.
- **Take it Further –** Have the students find the focal point of their Jell-O concave lens. They can do this by moving the plate farther away from the wall until the light converges into one single point on the wall. Then, measure the distance – this is the focal distance of their lens.

Discussion Questions

Lenses, pg. 197

1. What fact about light do lenses take advantage of? (*Lenses take advantage of the fact that light bends as it passes from air to glass.*)
2. How does the shape of the lens matter? (*The shape of the lens determines whether the light is bent towards or away from a focal point.*)

Optical Instruments, pg. 198

1. What do optical instruments do? (*Optical instruments magnify objects so that we can see more detail than we could with the naked eye.*)
2. How does a compound microscope work? (*A compound microscope magnifies objects in two different stages. In the first stage, light is reflected through the specimen up to an objective lens, which magnifies the object. In the second stage, the eye lens acts as a magnifying glass to increase the size of the image produced by the objective lens.*)
3. What is the difference between reflecting and refracting telescopes? (*Reflecting telescopes have a large concave mirror that collects light, with a second mirror that reflects the light to the eye lens. Refracting telescopes have a large convex mirror that refracts light to form on upside-down projection of the image, which is magnified by the eye lens.*)

Want More

- **Telescope** – Have the students make their own telescope using the directions from the following website:
  - http://www.space.com/24114-how-to-build-a-telescope-science-fair-projects.html

Sketch Week 20

Concave vs. Convex

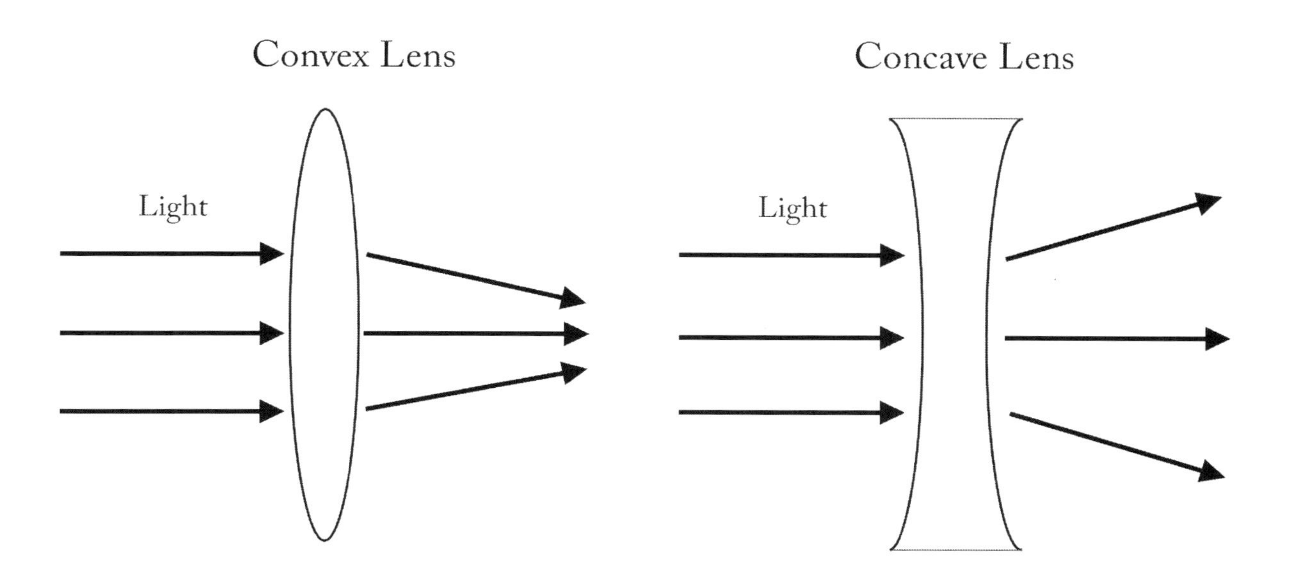

Unit 5 Light
Unit Test Answers

Vocabulary Matching

1. B
2. E
3. K
4. F
5. D
6. C
7. I
8. H
9. G
10. J
11. L
12. A

True or False

1. False (*Light can travel in a vacuum.*)
2. True
3. True
4. False (*When light travels from one transparent material to another, it bends. This is known as refraction.*)
5. True
6. False (*Colors are much clearer and brighter during the day than they are during the night.*)
7. True
8. True

Short Answer

1. Quantum theory suggests that light has a combination of the properties of a wave and of a particle.
2. The law of reflection states that the angle of reflection is equal to the angle of incidence.
3. The lens of the eye focuses the light on the retina, which has light-sensitive cells. These cells send signals through the optic nerve to the brain. The brain interprets the signals and builds a picture that we can see.
4. A compound microscope magnifies objects in two different stages. In the first stage, light is reflected through the specimen up to an objective lens, which magnifies the object. In the second stage, the eye lens acts as a magnifying glass to increase the size of the image produced by the objective lens.
5. Students should have included four of the following:
   - ✓ **Radio waves –** These waves are typically found in radios, televisions, microwaves, and radars.
   - ✓ **Infrared rays –** These waves are used as a source of heat as well as to find areas of heat differences.
   - ✓ **Visible light –** These waves make up the spectrum of visible light we can see.
   - ✓ **Ultraviolet rays –** These waves can kill microorganisms, produce vitamin D in our skin, and help plants grow.
   - ✓ **X-rays –** These waves have very short wavelengths and are used to make pictures of what is inside a solid object.
   - ✓ **Gamma rays –** These waves have the shortest wavelengths in the electromagnetic spectrum. Gamma rays are used in medicine to kill cancer cells and to provide images of the brain.

Unit 5 Light
Unit Test

Vocabulary Matching

1. Electromagnetic waves ____
2. Electromagnetic spectrum ____
3. Photons ____
4. Reflection ____
5. Refraction ____
6. Convex ____
7. Concave ____
8. Interference ____
9. Diffraction ____
10. Converging lens ____
11. Diverging lens ____
12. Lens ____

A. A lens that causes parallel light rays passing through it to spread out.

B. These waves carry energy from one place to another and are produced when an electric charge vibrates or accelerates.

C. A lens that curves outward.

D. The change in direction of a light beam as it passes from one medium to another of different density.

E. The full range of electromagnetic radiation, which includes radio waves, infrared rays, visible light, Ultraviolet light, X-rays, and Gamma rays.

F. The bouncing back of light from a surface.

G. The spreading out of light waves when they pass through a narrow slit.

H. The disturbance of a signal when two or more waves meet.

I. A lens that curves inward.

J. A piece of transparent material with a curved edge that causes light to bend in a particular way.

K. Packets of electromagnetic energy.

L. A lens that causes parallel light rays passing through it to come together.

True or False

1. ____________________ Light cannot travel in a vacuum.

2. ____________________ Electromagnetic waves all travel at the speed of light.

3. ____________________ Objects that don't make light reflect it so that we can see them.

4. ____________________ When light travels from one transparent material to another, it always continues on the same path.

5. ____________________ A color is formed by a wavelength or combination of wavelengths of light.

6. ____________________ We see colors the same at night as we do during the day.

7. ____________________ The shape of the lens determines whether the light is bent towards or away from a focal point.

8. ____________________ Optical instruments magnify objects so that we can see more detail than we could with the naked eye.

Short Answer

1. What does quantum theory say about light?

2. What is the law of reflection?

3. How do the eyes and brain work together to allow us to see?

4. How does a compound microscope work?

5. Name four of the six types of types of electromagnetic waves

Physics Unit 6

Electricity and Magnetism

Unit 6 Electricity and Magnetism

Overview of Study

Sequence of Study

Week 21: Electricity
Week 22: Conductors and Insulators
Week 23: Batteries
Week 24: Circuits
Week 25: Magnetism
Week 26: Electromagnetism
Week 27: Motors and Generators

Materials by Week

Week	Materials
21	Styrofoam pan, Aluminum pan, Wool, Plastic tongs
22	Light bulb, Copper wire, D battery, Electrical tape, Alligator clips, Organic material, such as a pickle, lemon slice, cheese, bread, or leaf
23	2 AA disposable batteries (one fully charged, one completely dead), Ruler
24	Computer with Internet connection
25	2 different types of magnets, such as a horseshoe magnet and a neodymium magnet, Paper clips (20 to 30), Paper, Cardboard, Thick books
26	D battery, Insulated copper wire – about 3 ft (1 m), 2 to 3 inch (5 to 8 cm) Nail, Electrical tape, Iron filings, Paper
27	Straws, Electrical tape, 6 ft. (2 m) of thin insulated wire, AAA battery, sandpaper, needle

Vocabulary for the Unit

1. **Electrical charge** – A charge produced by an excess or shortage of electrons.
2. **Induction** – The transfer of electrical charge without contact between the materials.
3. **Conductor** – A material through which electrical charge can easily flow.
4. **Insulator** – A material through which electrical charge cannot easily flow.
5. **Semiconductor** – A material that can act as a conductor or insulator based on its temperature.
6. **Anode** – A positively charged diode.
7. **Cathode** – A negatively charged diode.
8. **Diode** – An electrical component that allows electrical current to flow in and out in a single direction.
9. **Parallel circuit** – An electrical circuit in which current can pass though more than one path.
10. **Resistance** – The ability of a material to resist the flow of electrical current.
11. **Series circuit** – An electrical circuit in which current passes through components that are one after another.
12. **Electrolyte** – A substance that can conduct electrical current.

13. **Magnetic field** – The area around a magnet in which the magnetic force can be felt.
14. **Magnetic pole** – One of the two ends of a magnet where the force of attraction or repulsion is the strongest.
15. **Electromagnet** – A magnet that can be switched off and on with electric current.
16. **Electromagnetic force** – The force produced when an electrical current flows through a wire; it forms a magnetic field.
17. **Solenoid** – A coil of wire that behaves like a magnet when electric current passes through it.
18. **Turbine** – A machine with shafts and blades that are turned by the force of wind or steam; this movement generates energy that can be turned into electricity.

Memory Work for the Unit

Law of Electrostatics

Like charges repel each other and opposite charges attract each other.

Types of Electrical Current

1. **DC** – Direct Current is when the electrical charge only flows in one direction.
2. **AC** – Alternating Current is when the electrical charge regularly reverses the direction of its flow.

First Law of Magnetism

Like poles repel, while unlike poles attract.

Equations

- Electrical Charge Equation

  $Q = I \cdot t$

  "Q" stands for electrical charge (measured in coulombs).
  "I" stands for current (measured in amps).
  "t" stands for time.

- Potential Difference Equation

  $V = \frac{E}{C}$

  "V" stands for potential difference or voltage (measured in volts).
  "E" stands for energy transferred (measured in joules).
  "C" stands for electrical charge (measured in coulombs).

- Ohm's Law

  $V = I \cdot R$

  "V" stands for voltage (measured in volts).
  "I" stands for current (measured in amps).
  "R" stands for resistance (measured in ohms).

Notes

Student Assignment Sheet Week 21
Electricity

Experiment: Can I transfer an electrical charge?

Materials
- ✓ Styrofoam tray or plate
- ✓ Aluminum pan or pie-plate
- ✓ Wool
- ✓ Plastic tongs
- ✓ Pin

Procedure
1. Read the introduction to the experiment and answer the question for the hypothesis section.
2. Rub the Styrofoam tray (or plate) vigorously with a piece of wool for about two minutes. Set the tray face down on a smooth surface.
3. Use the plastic tongs to set the aluminum pan (or pie-plate) on top of the Styrofoam tray (or plate). Then, gently slide it back and forth several times.
4. Now, use the tongs to pick up the pin. With your hands on the plastic tongs, move the pin close to the pan and observe what happens.
5. Draw conclusions and complete the experiment sheet.

Vocabulary & Memory Work
- ❒ Vocabulary: electrical charge, induction
- ❒ Memory Work—This week, work on memorizing the law of electrostatics and the electrical charge equation.
  - Law of Electrostatics – Like charges repel each other and opposite charges attract each other.
  - Electrical charge (C) = Current (I) • time (t)

Sketch: Anatomy of a Lightning Strike
- Label the following – negative charges collect at the bottom of a storm clouds, positive charges collect on surface of the ground, electrical discharge path

Writing
- Reading Assignment: *DK Encyclopedia of Science* pg. 145 Electricity and Magnetism, pp. 146-147 Static Electricity
- Additional Research Readings
  - Electricity: *KSE* pp. 338-339
  - Static Electricity: *UIDS* pp. 56-57

Dates
- 585 BC – Greek philosopher, Thales, inadvertently discovers static electricity when he rubs a piece of amber with fur and observes how the amber now attracts small objects, like feathers.
- 1544-1603 – William Gilbert lives. He is known as the father of electricity and magnetism.
- 1753 – Benjamin Franklin announces his lightning conductor invention, which he created as a result of his famous kite-flying experiments.
- 1784-1789 – Charles Coulomb, a French physicist, writes and proves the law of electrostatics.

Schedules for Week 21

Two Days a Week

Day 1	Day 2
☐ Do the "Can I transfer an electrical charge?" experiment, and then fill out the experiment sheet on SG pp. 160-161 ☐ Define electrical charge and induction on SG pg. 156 ☐ Enter the dates onto the date sheets on SG pp. 8-13	☐ Read pp. 145-147 from *DK EOS,* and then discuss what was read ☐ Color and label the "Anatomy of Lightning Strike" sketch on SG pg. 159 ☐ Prepare an outline or narrative summary, write it on SG pp. 162-163
Supplies I Need for the Week ✓ Styrofoam tray or plate ✓ Aluminum pan or pie-plate ✓ Wool, Plastic tongs, Pin	
Things I Need to Prepare	

Five Days a Week

Day 1	Day 2	Day 3	Day 4	Day 5
☐ Do the "Can I transfer an electrical charge?" experiment, and then fill out the experiment sheet on SG pp. 160-161 ☐ Enter the dates onto the date sheets on SG pp. 8-13	☐ Read pp. 145-147 from *DK EOS,* and then discuss what was read ☐ Write an outline on SG pg. 162	☐ Define electrical charge and induction on SG pg. 156 ☐ Color and label the "Anatomy of Lightning Strike" sketch on SG pg. 159	☐ Read one or all of the additional reading assignments ☐ Write a report on what you learned on SG pg. 163	☐ Complete one of the Want More Activities listed **OR** ☐ Study a scientist from the field of Physics
Supplies I Need for the Week ✓ Styrofoam tray or plate ✓ Aluminum pan or pie-plate ✓ Wool, Plastic tongs, Pin				
Things I Need to Prepare				

Additional Information Week 21

Experiment Information

- **Introduction** – (*from the Student Guide*) An electrical charge is produced when a subatomic particle within an object has a shortage or excess of electrons. A shortage of electrons produces a positive electrical charge and an excess of electrons produces a negative electrical charge. The two types of electrical charges are attracted to one another and electricity takes advantage of this principle. In today's experiment, you are going to test whether you can transfer an electrical charge from one object to another.
- **Results –** The students should hear and see a spark as the pin gets within a half-an-inch or less from the pan.
- **Explanation –** By rubbing the Styrofoam tray with the wool, the students produced a static electrical charge through friction. This static electric charge is then transferred to the aluminum pan through contact. When the pin came in close contact to the aluminum pan, the static charge was transferred by induction from the pan to the pin. This process gave the students a firsthand look at the three methods of transferring electrical charge – friction, contact, and induction.
- **Troubleshooting –** If the students did not see a spark, you may need to rub the Styrofoam more vigorously with the wool. You can also have the students try rubbing the tray against their hair.
- **Take it Further –** Have the students repeat the experiment using metal objects of different thicknesses, such as a paper clip or a nail, to see if they get a larger spark.

Discussion Questions

Electricity and Magnetism, pg. 145

1. What is electronics? (*Electronics is the use of components or materials to control electricity in a way that is useful for during work.*)
2. What is lodestone and what was it used for? (*Lodestone is a mineral that is naturally magnetized. In the early days, it was used by sailors as a compass.*)

Static Electricity, pp. 146-147

1. What is static electricity? (*Static electricity is a stationary electrical charge.*)
2. What happens when you charge an object through friction? (*When you charge an object through friction, the electrons from one object are rubbed off onto another object, creating a static charge.*)
3. What is attraction? (*Attraction is when two objects with opposite electrical charges move towards one another.*) Repulsion? (*Repulsion is when two objects with the same electrical charge move away from one another.*)
4. What is a capacitor? (*A capacitor is a device used to store electrical charge within a piece of electronic equipment.*)

Want More

- **Electroscope –** Have the students make their own electroscope using the directions from the following website:
  - http://www.education.com/science-fair/article/physics_making-electroscope/

- **Electrical Charge** – Have the students test the effect of electrical charge on a stream of water. Begin by having the students rub a plastic comb with wool. Then, place the comb near a stream of water and observe what happens. (*The students should see that the stream of water is attracted to the comb. This is because the comb has been charged by the wool and it now can attract the slightly polar stream of water.*)
- **Electrical Charge Worksheet** – Have the students practice using the equation to calculate the electrical charge of a system with the worksheet in the Appendix on pg. 260.
  Answers
  1. 5400 C
  2. 36,000 C
  3. 68 amps
  4. 13.4 sec

Sketch Week 21

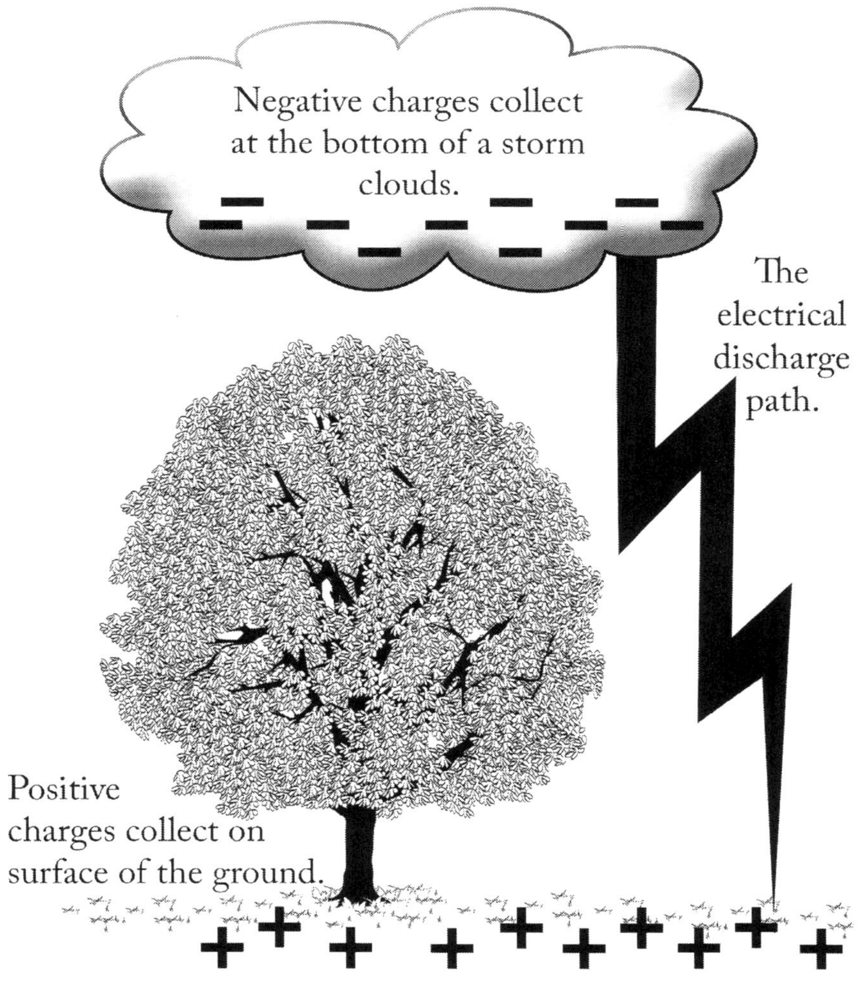

Student Assignment Sheet Week 22
Conductors and Insulators

Experiment: Can organic materials conduct electricity?

Materials
- ✓ Light bulb
- ✓ Copper wire
- ✓ D battery
- ✓ Electrical tape
- ✓ 2 Alligator clips
- ✓ Organic material, such as a pickle, lemon slice, cheese, bread, or a leaf

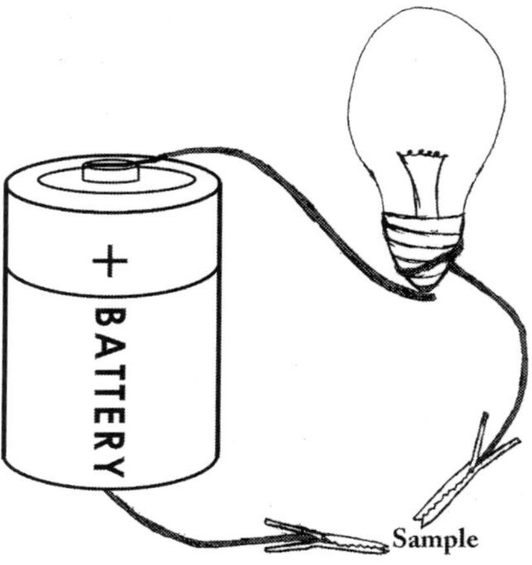

Procedure
1. Read the introduction to the experiment and answer the question for the hypothesis section.
2. Begin by cutting the wire into three lengths. Then, take one of the wires and attach an alligator clip to one end. Take the next wire and attach one end to the other alligator clip. Then, wrap the other end once around the base of a light bulb.
3. Now, take the two wires with bare ends and attach those to the two terminals of the battery using the electrical tape. (*See the diagram above to visually check your electrical circuit.*)
4. The electrical circuit is now ready for testing. Simply clip the alligator clips on either side of your sample, hold the top of the bulb, and touch the base of the light bulb to the end of the wire coming from the battery to complete the circuit. If the light bulb lights up, the material conducts electricity; if it does not, the sample does not conduct electricity.
5. Draw conclusions and complete the experiment sheet.

Vocabulary & Memory Work
- ☐ Vocabulary: conductor, insulator, semiconductor
- ☐ Memory Work—This week, begin working on memorizing the two types of electrical current and the potential difference equation. (*See Unit Overview Sheet for a complete listing.*)

Sketch: Electron Flow Diagram
- Label the following – positive terminal, negative terminal, electron flow, current flow
- Draw the arrows to show the direction of the electron flow and current.

Writing
- Reading Assignment: *DK Encyclopedia of Science* pp. 148-149 (Current Electricity)
- Additional Research Readings
  - Conductors: *KSE* pp. 360-361
  - Insulators: *KSE* pp. 362-363
  - Semiconductors: *UIDS* pg. 65
  - Electrical Current: *UIDS* pp. 60-61

Dates
- 1987 – The Nobel Prize is awarded to Muller and Bednorz for their work with finding superconductors that function above absolute zero.

Schedules for Week 22

Two Days a Week

Day 1	Day 2
☐ Do the "Can organic materials conduct electricity?" experiment, and then fill out the experiment sheet on SG pp. 166-167 ☐ Define conductor, insulator, and semiconductor on SG pg. 156 ☐ Enter the dates onto the date sheets on SG pp. 8-13	☐ Read pp. 148-149 from *DK EOS,* and then discuss what was read ☐ Color and label the "Electron Flow Diagram" sketch on SG pg. 165 ☐ Prepare an outline or narrative summary; write it on SG pp. 168-169

Supplies I Need for the Week

- ✓ Light bulb, Copper wire, D battery
- ✓ Electrical tape, 2 Alligator clips
- ✓ Organic material, such as a pickle, lemon slice, cheese, bread, or a leaf

Things I Need to Prepare

Five Days a Week

Day 1	Day 2	Day 3	Day 4	Day 5
☐ Do the "Can organic materials conduct electricity?" experiment, and then fill out the experiment sheet on SG pp. 166-167 ☐ Enter the dates onto the date sheets on SG pp. 8-13	☐ Read pp. 148-149 from *DK EOS,* and then discuss what was read ☐ Write an outline on SG pg. 168	☐ Define conductor, insulator, and semiconductor on SG pg. 156 ☐ Color and label the "Electron Flow Diagram" sketch on SG pg. 165	☐ Read one or all of the additional reading assignments ☐ Write a report on what you learned on SG pg. 169	☐ Complete one of the Want More Activities listed **OR** ☐ Study a scientist from the field of Physics

Supplies I Need for the Week

- ✓ Light bulb, Copper wire, D battery
- ✓ Electrical tape, 2 Alligator clips
- ✓ Organic material, such as a pickle, lemon slice, cheese, bread, or a leaf

Things I Need to Prepare

Additional Information Week 22

Experiment Information

- ☞ **Introduction** – (*from the Student Guide*) A conductor is a material through which electrical charge can easily flow, while an insulator is a material through which electrical charge cannot easily flow. When we want to move electricity from one place to another, we use copper wire, which easily conducts electrical charge. But what if we could use other renewable organic materials, like pickles, to conduct electricity? In today's lab, you are going to test whether different organic compounds can conduct electricity.
- ☞ **Results** – The students' results will vary based on the materials they choose to test, but they will see that several organic materials, such as the pickle, lemon slice, and leaf, all conduct electricity.
- ☞ **Explanation** – Organic compounds that are slightly acidic, such as pickles and lemons, are able to conduct electricity. This is due to the ions present in an acidic solution. Other organic materials, like plants, are able to conduct electricity due to the slight magnetic field that exist within them. For a material to be a conductor, the electrical charge needs to move freely through the substance and the presence of ions allows the charge to do just that. Of course, using pickles to move electricity throughout our homes would be inefficient, considering how often we would have to change them.
- ☞ **Troubleshooting** – If the students are having trouble getting the wires to wrap around the bulb and stay put, you may want to try making your own socket using a rubber band and two tongue depressors. Directions for this can be found at the following website:
  - 💻 http://www.instructables.com/id/Flashlight-Bulb-Socket-for-EXPERIMENTS/
- ☞ **Take it Further** – Have the students play the conductors and insulators game at the following website:
  - 💻 http://www.helcohi.com/sse/wires/game.html

Discussion Questions

1. What is electric current and how is it measured? (*Electric current is the flow of electricity from one place to another. We use amperes or amps to measure electric current.*)
2. What is the difference between a conductor and an insulator? (*A conductor allows current to pass through, but an insulator does not.*)
3. What does it mean to "dope" a semiconductor and why would you do that? (*To "dope" a semiconductor means to make it impure. This is done in electronics to make either a negative or positive semiconductor, which allows for the electric current to be carried through a device.*)
4. How does electroplating work? (*In electroplating, a current runs through the solution and the material, such as circuit boards. The material acts as an anode, which attracts the positive metal ions, causing the metal ions to attach to the material, thereby creating a thin layer of the metal on the material.*)

Want More

- ✐ **Semiconductors** – Have the students research about semiconductors and write a brief report about their use in modern electronics.

- **Potential Difference Worksheet** – Have the students practice using the potential difference equation to calculate the volts generated by a system with the worksheet in the Appendix on pg. 261.

  Answers:
  1. 15 volts
  2. 20 C
  3. 78 J

Sketch Week 22

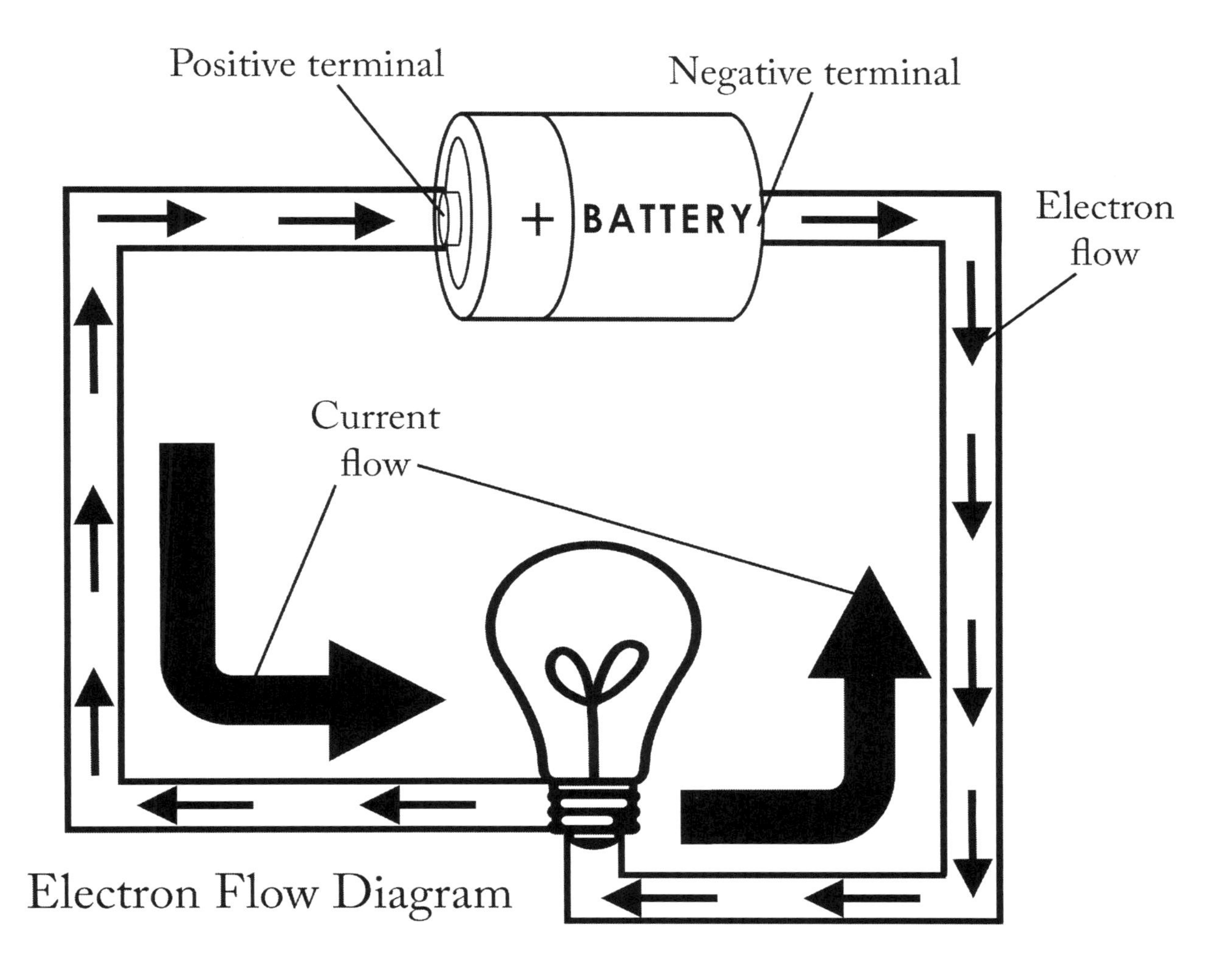

Student Assignment Sheet Week 23
Batteries

Experiment: Do dead batteries bounce?

Materials
- ✓ 2 AA disposable batteries (one fully charged, one completely dead)
- ✓ Ruler

Procedure
1. Read the introduction to the experiment and answer the question for the hypothesis section.
2. Drop the fully charged battery from a height of 3 inches (7.5 cm). Observe and measure the height of any bounce that occurs. Record the results on your experiment sheet.
3. Drop the completely dead battery from a height of 3 inches (7.5 cm). Observe and measure the height of any bounce that occurs. Record the results on your experiment sheet.
4. Draw conclusions and complete the experiment sheet.

Vocabulary & Memory Work
- ☐ Vocabulary: anode, cathode, diode
- ☐ Memory Work—This week, continue to work on memorizing the law of electrostatics and the two types of electrical current.

Sketch: Anatomy of a dry cell battery
- Label the following – positive terminal, negative terminal, Zinc casing (negative electrode), Carbon rod (positive electrode), powdered carbon and magnesium oxide, ammonium chloride paste (electrolyte)

Writing
- Reading Assignment: *DK Encyclopedia of Science* pp. 150-151 (Cells and batteries)
- Additional Research Readings
  - Electrochemistry: *KSE* pg. 356
  - Power Cells: *KSE* pg. 357
  - Cells and batteries: *UIDS* pp. 68-69

Dates
- 1800 – Italian scientist, Volta, invents the first battery able to hold an electrical charge.

Schedules for Week 23

Two Days a Week

Day 1	Day 2
☐ Do the "Do dead batteries bounce?" experiment, and then fill out the experiment sheet on SG pp. 172-173 ☐ Define anode, cathode, and diode on SG pg. 156 ☐ Enter the dates onto the date sheets on SG pp. 8-13	☐ Read pp. 150-151 from *DK EOS,* and then discuss what was read ☐ Color and label the "Anatomy of a dry cell battery" sketch on SG pg. 171 ☐ Prepare an outline or narrative summary; write it on SG pp. 174-175

Supplies I Need for the Week

- ✓ 2 AA disposable batteries (one fully charged, one completely dead)
- ✓ Ruler

Things I Need to Prepare

Five Days a Week

Day 1	Day 2	Day 3	Day 4	Day 5
☐ Do the "Do dead batteries bounce?" experiment, and then fill out the experiment sheet on SG pp. 172-173 ☐ Enter the dates onto the date sheets on SG pp. 8-13	☐ Read pp. 150-151 from *DK EOS,* and then discuss what was read ☐ Write an outline on SG pg. 174	☐ Define anode, cathode, and diode on SG pg. 156 ☐ Color and label the "Anatomy of a dry cell battery" sketch on SG pg. 171	☐ Read one or all of the additional reading assignments ☐ Write a report on what you learned on SG pg. 175	☐ Complete one of the Want More Activities listed **OR** ☐ Study a scientist from the field of Physics

Supplies I Need for the Week

- ✓ 2 AA disposable batteries (one fully charged, one completely dead)
- ✓ Ruler

Things I Need to Prepare

Additional Information Week 23

Experiment Information

- **Introduction** – (*from the Student Guide*) Batteries hold electrical charge that is used to power many of our modern-day portable electronics. For years, there has been an old-wives' tale that you can test if a battery is dead by dropping it. If it bounces, it is dead. If not, it still holds a charge. In today's experiment, you are going to test to see if this myth is true.
- **Results** – The students should see that the completely dead battery does bounce.
- **Explanation** – Disposable batteries have two chambers, a positively-charged cathode filled with manganese dioxide, and a negatively-charged anode filled with zinc-containing gel and potassium hydroxide. When the two ends are connected, a chemical reaction occurs with the zinc, which allows the current to move to the cathode, creating electricity. This reaction happens until all the zinc is used up and the battery goes dead. As the zinc is used up, the gel physically changes into a pocket-filled ceramic-like material that acts as a spring. This physical change gives the battery the ability to bounce as it goes dead.
- **Take it Further** – Have the students repeat the experiment with batteries that have a varying degree of discharge, i.e., one at 75%, one at 50%, and one at 25% discharge. (*The students should see that the more the battery has been discharged, the higher it will bounce. However after about 50% discharge, the change in the height of the bounce levels off.*)

Discussion Questions

1. What do all batteries have in common? (*All batteries store chemical energy and change it into electrical energy.*)
2. What is the difference between a cell and a battery? (*A cell is the basic unit that can produce electricity. A battery has two or more cells.*)
3. What are the three basic components of a cell? (*The three basic components of a cell are the positive electrode, the negative electrode, and an electrolyte chemical.*)
4. What is the benefit of a larger cell battery? (*A larger cell battery produces the same electromotive force, but it lasts longer than a smaller cell.*)
5. How does a solar cell work? (*Solar cells change light energy into electricity instead of chemical energy.*)

Want More

- **Battery Brands** – Have the students test to see which brand of battery lasts the longest. You can use three different battery brands and a flashlight to test this. Have the students load each brand of battery into the flashlight, turn it on, and observe how long it takes for the light to go out.
- **Lemon battery** – Have the students build their own battery using a lemon. Watch the following video for directions:
    - https://www.youtube.com/watch?v=_DPIMeY_z1w

Sketch Week 23

Anatomy of a Dry Cell Battery

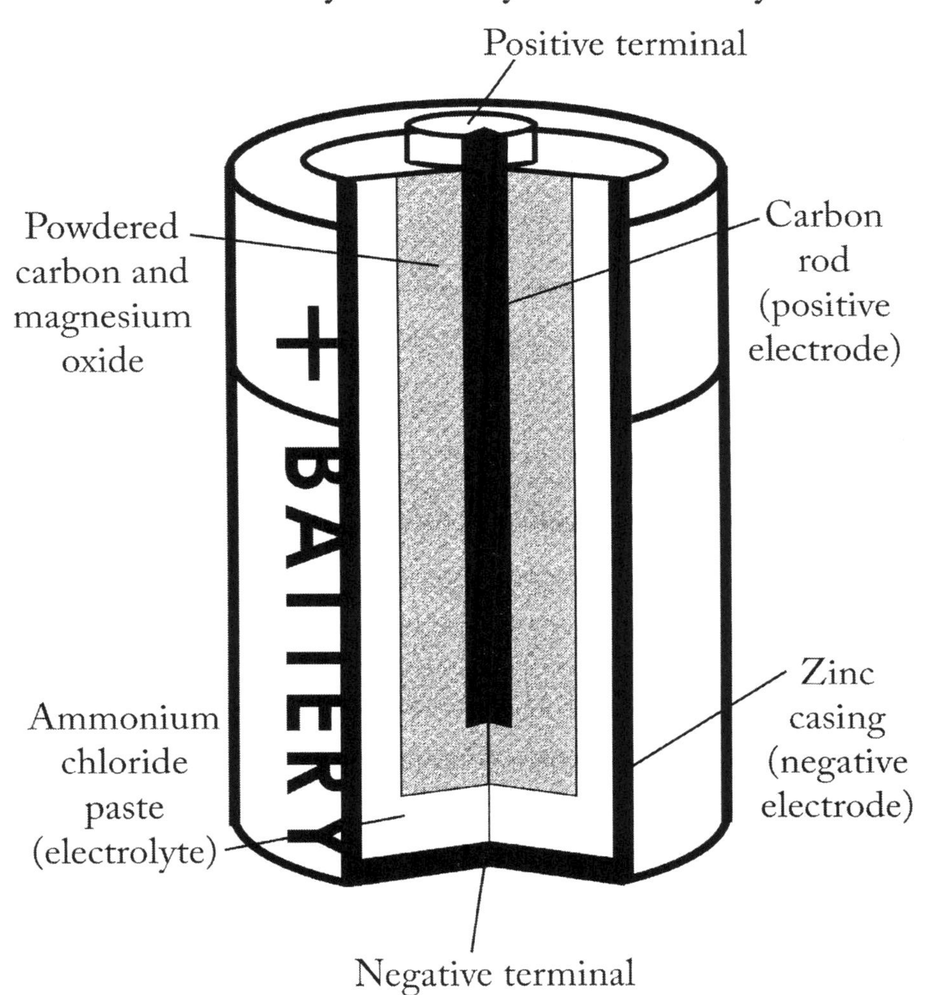

Student Assignment Sheet Week 24
Circuits

Experiment: Online Circuits Lab

Materials

- ✓ Computer with Internet connection

Procedure

1. Before you begin the lab, draw a circuit that you think will work under the hypothesis section on your experiment sheet.
2. Begin the online lab by clicking on the website below to begin your lab:
   http://phet.colorado.edu/en/simulation/circuit-construction-kit-dc-virtual-lab
   Note—If you cannot get the link to work, head to PhET's main website at http://phet.colorado.edu/. Enter "circuits" into the search box and then click on the link for the "Circuit Construction Kit (DC Only)."
3. Click on the "Run Now" button and then use the available materials in the box to test the circuit you designed. Record on your experiment sheet whether the proposed circuit worked.
4. Now, use the materials to create two of three different types of circuits.
5. Draw two of these circuits you created on your experiment sheet and write down what you have learned.

Vocabulary & Memory Work

- Vocabulary: parallel circuit, resistance, series circuit
- Memory Work—This week, work on memorizing the Ohm's Law.
  - Voltage (V) = Current (I) • Resistance (R)

Sketch: Circuit Diagram

- Label the following – Battery, resistor, switch, light
- Draw a circuit using the symbols on the sketch sheet that depict a pair of bulbs in series and a set of parallel-connected pair of bulbs.

Writing

- Reading Assignment: *DK Encyclopedia of Science* pp. 152-153 (Circuits)
- Additional Research Readings
  - Electrical Circuits: *KSE* pp. 340-341
  - Resistance: *KSE* pg. 363
  - Controlling Current: *UIDS* pp. 62-63

Dates

- 1827 – George Ohm publishes his work on resistance, including an equation that eventually becomes known as Ohm's Law.
- 1828 – Andre Ampere is elected as a member of the Royal Swedish Academy of Science in recognition of his contributions to the creation of the field of modern electrical science.

Schedules for Week 24

Two Days a Week

Day 1	Day 2
☐ Do the "Online Circuit Lab" experiment, and then fill out the experiment sheet on SG pp. 178-179 ☐ Define parallel circuit, resistance, and series circuit on SG pg. 156-157 ☐ Enter the dates onto the date sheets on SG pp. 8-13	☐ Read pp. 152-153 from *DK EOS,* and then discuss what was read ☐ Color and label the "Circuit Diagram" sketch on SG pg. 177 ☐ Prepare an outline or narrative summary; write it on SG pp. 180-181
Supplies I Need for the Week ✓ Computer with Internet connection	
Things I Need to Prepare	

Five Days a Week

Day 1	Day 2	Day 3	Day 4	Day 5
☐ Do the "Online Circuit Lab" experiment, and then fill out the experiment sheet on SG pp. 178-179 ☐ Enter the dates onto the date sheets on SG pp. 8-13	☐ Read pp. 152-153 from *DK EOS,* and then discuss what was read ☐ Write an outline on SG pg. 180	☐ Define parallel circuit, resistance, and series circuit on SG pg. 156-157 ☐ Color and label the "Circuit Diagram" sketch on SG pg. 177	☐ Read one or all of the additional reading assignments ☐ Write a report on what you learned on SG pg. 181	☐ Complete one of the Want More Activities listed **OR** ☐ Study a scientist from the field of Physics
Supplies I Need for the Week ✓ Computer with Internet connection				
Things I Need to Prepare				

Additional Information Week 24

Experiment Information

☞ **Troubleshooting –** If you cannot get the file to work, you may need to update your Java reader before playing the game.

☞ **Take it Further –** Have the student choose one of the other demonstrations to do. The "Circuit Construction Kit (AC + DC)", "Ohm's Law", and "Semiconductors" are good options.

Discussion Questions

1. What is a circuit? (*A circuit is the path that electricity takes when it flows.*)
2. How does a flashlight represent a simple circuit? (*In a flashlight, electricity flows from the battery through the switch, through the bulb, and back to the battery.*)
3. How does resistance affect the flow of current in a circuit? (*The greater the resistance in a circuit, the less the current flows.*)
4. How is resistance affected when the resistors are connected in series? (*When resistors are connected in series, the overall resistance in a circuit increases.*) In parallel? (*When resistors are connected in parallel, the overall resistance in a circuit decreases.*)

Want More

✐ **Circuit Building –** Purchase a kit, like Snap Circuits, and have the students build some of the circuits from the kit.

✐ **Ohm's Law Worksheet –** Have the students practice using the resistance equation (also known as Ohm's Law) to calculate the volts generated by a system with the worksheet in the Appendix on pg. 262.

Answers:

1. 36 volts
2. 1.4 amps
3. 3 amps
4. 90 volts

Sketch Week 24

The circuit below is just one of many options that the students could have drawn.

Circuit Diagram

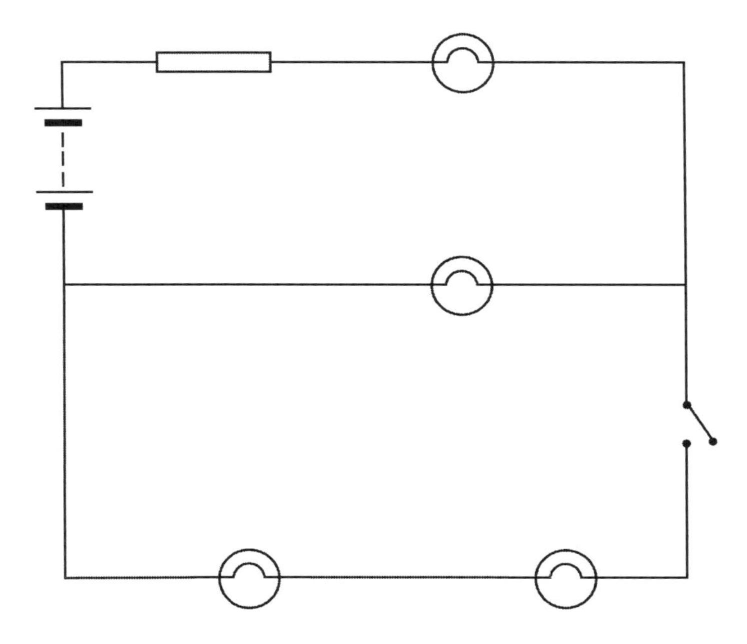

Symbols

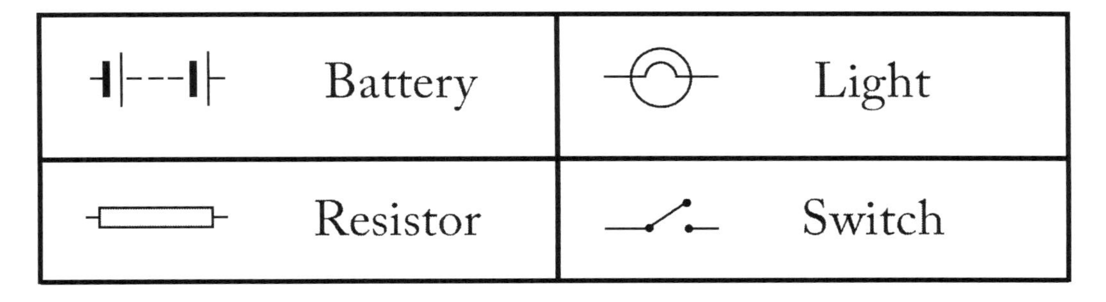

Symbol	Name	Symbol	Name
	Battery		Light
	Resistor		Switch

Student Assignment Sheet Week 25
Magnetism

Experiment: Do different types of magnets have different strengths?

Materials
- ✓ 2 different types of magnets, such as a horseshoe magnet and a neodymium magnet
- ✓ Paper clips (20 to 30)
- ✓ Paper
- ✓ Cardboard
- ✓ Thick books

Procedure
1. Read the introduction to the experiment and answer the question for the hypothesis section.
2. Use the following tests to determine the strength of each of your magnets:
    - **Paper clip test** – Pick up one paper clip with the magnet. Then attach another one to the bottom of the first, creating a chain of paper clips. Continue to add more paper clips until the chain breaks or you cannot add anymore. Count how many paper clips you added and record that on your experiment sheet.
    - **Thickness test** – Have a partner hold a piece of paper with a paper clip on it. Use the magnet to move the paper clip around. Next, switch the paper for a piece of cardboard and repeat. Then, switch the cardboard for a thick book. Keep adding books until the magnet can no longer attract the paper clip and move it. Measure the thickness of the material that the magnet could attract through and record that on your experiment sheet.
3. Draw conclusions and complete the experiment sheet.

Vocabulary & Memory Work
- Vocabulary: electrolyte, magnetic field, magnetic pole
- Memory Work—This week, work on memorizing the first law of magnetism.
    - **First law of magnetism** – Like poles repel, while unlike poles attract.

Sketch: Magnetic Attraction and Repulsion
- Draw the magnetic fields that show the repulsive force on one set of magnets and the attractive force on the other set.
- Label the following – south pole, north pole, repulsive forces, attractive forces

Writing
- Reading Assignment: *DK Encyclopedia of Science* pp. 154-155 (Magnetism)
- Additional Research Readings
    - Magnets and Magnetism: *KSE* pp. 342-343
    - Magnets: *UIDS* pp. 70-71
    - Magnetic Fields: *UIDS* pp. 72-73

Dates
- 1100 – Chinese sailors are the first on record for using a magnetic compass for navigating on a cloudy day.
- 1600 – William Gilbert, an English doctor and physicist, publishes a book which for the first time explains exactly how a compass works.
- 1952 – Magnets are made out of ceramics for the first time.
- 1983 – Neodymium magnets are first invented.

Schedules for Week 25

Two Days a Week

Day 1	Day 2
☐ Do the "Do different types of magnets have different strengths?" experiment, and then fill out the experiment sheet on SG pp. 184-185 ☐ Define electrolyte, magnetic field, and magnetic pole on SG pg. 157 ☐ Enter the dates onto the date sheets on SG pp. 8-13	☐ Read pp. 154-155 from *DK EOS,* and then discuss what was read ☐ Color and label the "Magnetic Attraction and Repulsion" sketch on SG pg. 183 ☐ Prepare an outline or narrative summary; write it on SG pp. 186-187
Supplies I Need for the Week ✓ 2 different types of magnets, such as a horseshoe magnet and a neodymium magnet ✓ Paper clips (20 to 30) ✓ Paper, Cardboard, Thick books	
Things I Need to Prepare	

Five Days a Week

Day 1	Day 2	Day 3	Day 4	Day 5
☐ Do the "Do different types of magnets have different strengths?" experiment, and then fill out the experiment sheet on SG pp. 184-185 ☐ Enter the dates onto the date sheets on SG pp. 8-13	☐ Read pp. 154-155 from *DK EOS,* and then discuss what was read ☐ Write an outline on SG pg. 186	☐ Define electrolyte, magnetic field, and magnetic pole on SG pg. 157 ☐ Color and label the "Magnetic Attraction and Repulsion" sketch on SG pg. 183	☐ Read one or all of the additional reading assignments ☐ Write a report on what you learned on SG pg. 187	☐ Complete one of the Want More Activities listed **OR** ☐ Study a scientist from the field of Physics
Supplies I Need for the Week ✓ 2 different types of magnets, such as a horseshoe magnet and a neodymium magnet ✓ Paper clips (20 to 30) ✓ Paper, Cardboard, Thick books				
Things I Need to Prepare				

Additional Information Week 25

Experiment Information

- **Introduction –** (*from the Student Guide*) Magnets come in all shapes and sizes, like bars, horseshoes, discs, or rings. Three key metals have the ability to become permanently magnetized – cobalt, iron, and nickel. This property is known as ferromagnetism and it is the key to making magnets. In today's experiment, you are going to see if all magnets have the same strength or if their attractive force is dependent upon the materials of which they are composed.
- **Results –** The results will vary based on the size and composition of your magnets, but generally the neodymium magnet will surpass the standard iron-steel magnet.
- **Explanation –** The strength of a magnet's attractive forces depends upon its size and its chemical composition. Ceramic magnets, such as the horseshoe or bar magnet, are made by pressing and baking either barium ferrite or strontium ferrite powder into a shape. This shape is then cooled and magnetized to form a ceramic magnet. These magnets are relatively inexpensive and resist corrosion well, but ceramic magnets are weak. Neodymium magnets are made from a metal alloy mixture of neodymium, iron, and boron. These magnets corrode quickly, but they are 10 times stronger than the same-sized ceramic magnet.
- **Take it Further –** Have the students repeat the magnet strength tests using different sizes of the same type of magnet.

Discussion Questions

1. What do all magnets have? (*All magnets have a north and south pole.*)
2. What causes the Earth's magnetic field? (*The Earth's magnetic field is cause by the iron core at the center of the Earth.*) How does the magnetic field affect life on the surface of the Earth? (*The Earth's magnetic field allows us to use a compass to navigate. It also causes an aurora to be produced, known as the northern lights.*)
3. What causes a piece of steel to magnetize? (*A piece of steel will magnetize when you stroke it with a magnet, causing all the domains in the bar to align in the same direction.*)
4. What can happen with you hit a steel magnet with a hammer? *(Hitting a steel magnet with a hammer can cause the domains to shake up and the steel will lose its magnetism.)*
5. What is the purpose of a keeper on a horseshoe magnet? (*The keeper bar on a horseshoe magnet causes the domains in the magnet to remain aligned and prevents the magnet from losing its magnetism.*)

Want More

- **Magnetic fields –** Have the students use two bar-shaped magnets and iron filings to reproduce the magnetic fields shown on pg. 154 of the *DK Encyclopedia of Science*. They can also use the horseshoe and disc magnets from the experiment to see what kind of magnetic field the magnets create in the iron filings.
- **Earth magnetic field –** Have the students learn more about Earth's magnetic field. Then, have the students draw a poster depicting the magnetic field of the Earth.

Sketch Week 25

Magnetic Attraction and Repulsion

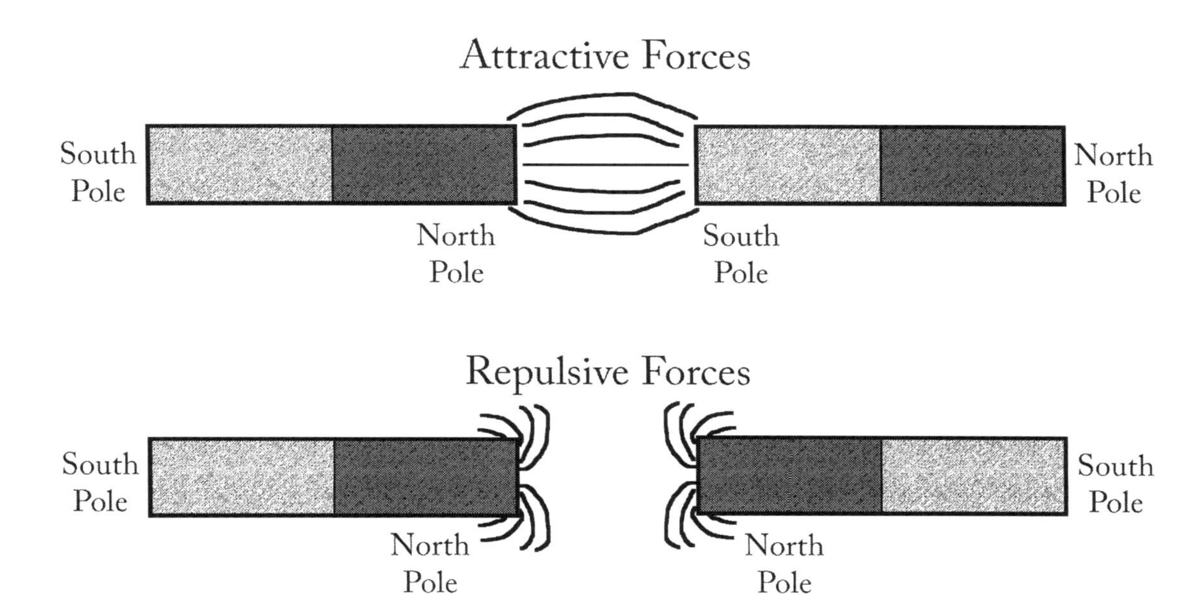

Student Assignment Sheet Week 26
Electromagnetism

Experiment: Can I create a magnetic field with electricity?

Materials
- ✓ D battery
- ✓ Insulated copper wire – about 3 ft (1 m)
- ✓ 2 to 3 in (5 to 8 cm) Nail
- ✓ Electrical tape
- ✓ Iron filings
- ✓ Paper

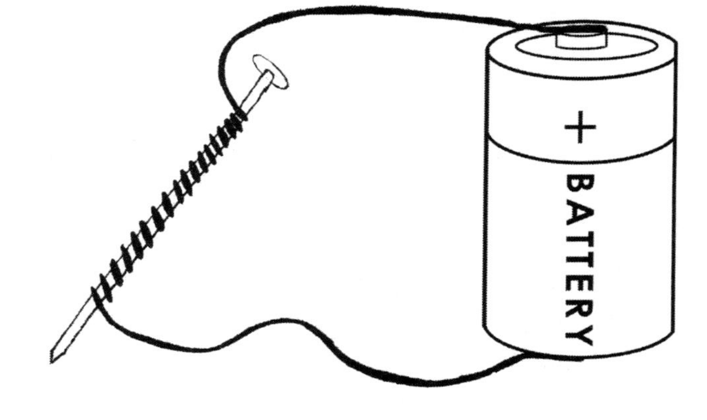

Procedure
1. Read the introduction to the experiment and answer the question for the hypothesis section.
2. Wrap the nail with the insulated wire in tight coils, leave at least a 6 in (16 cm) tail on each end. Using the tape, attach one end of the wire to the positive terminal on the battery and the other end to the negative terminal. (*See the diagram above to visually check your set-up.*)
3. Sprinkle some of the iron filings on a piece of paper. Take hold of the battery in the center and move the wire-wrapped nail close to the filings and observe what happens. If you see a magnetic field form in the iron filings, draw the field on your experiment sheet.
4. Draw conclusions and complete the experiment sheet.

Vocabulary & Memory Work
- ☐ Vocabulary: electromagnet, electromagnetic field
- ☐ Memory Work—This week, continue to work on memorizing the first law of magnetism.

Sketch: Magnetic Fields in an Electromagnet
- Draw several arrows show the flow of the electrical current in the circuit and the magnetic field created by the electrical current.
- Label the following – south pole, north pole

Writing
- Reading Assignment: *DK Encyclopedia of Science* pp. pp. 156-157 (Electromagnetism)
- Additional Research Readings
  - Electromagnetism: *KSE* pp. 344-345, *UIDS* pp.74-75

Dates
- 1820 – Hans Christiaan Oersted is performing an experiment with electrical current when he noticed that the needle of a nearby compass moved. This realization led to the discovery of electromagnetism.

Schedules for Week 26

Two Days a Week

Day 1	Day 2
☐ Do the "Can I create a magnetic field with electricity?" experiment, and then fill out the experiment sheet on SG pp. 190-191 ☐ Define electromagnet and electromagnetic field on SG pg. 157 ☐ Enter the dates onto the date sheets on SG pp. 8-13	☐ Read pp. 156-157 from *DK EOS,* and then discuss what was read ☐ Color and label the "Magnetic Fields in an Electromagnet" sketch on SG pg. 189 ☐ Prepare an outline or narrative summary; write it on SG pp. 192-193
Supplies I Need for the Week ✓ D battery, Insulated copper wire – about 3 ft (1 m) ✓ 2 to 3 in (5 to 8 cm) Nail, Electrical tape ✓ Iron filings, Paper	
Things I Need to Prepare	

Five Days a Week

Day 1	Day 2	Day 3	Day 4	Day 5
☐ Do the "Can I create a magnetic field with electricity?" experiment, and then fill out the experiment sheet on SG pp. 190-191 ☐ Enter the dates onto the date sheets on SG pp. 8-13	☐ Read pp. 156-157 from *DK EOS,* and then discuss what was read ☐ Write an outline on SG pg. 192	☐ Define electromagnet and electromagnetic field on SG pg. 157 ☐ Color and label the "Magnetic Fields in an Electromagnet" sketch on SG pg. 189	☐ Read one or all of the additional reading assignments ☐ Write a report on what you learned on SG pg. 193	☐ Complete one of the Want More Activities listed **OR** ☐ Study a scientist from the field of Physics
Supplies I Need for the Week ✓ D battery, Insulated copper wire – about 3 ft (1 m) ✓ 2 to 3 in (5 to 8 cm) Nail, Electrical tape ✓ Iron filings, Paper				
Things I Need to Prepare				

Additional Information Week 26

Experiment Information

- ☞ **Introduction –** (*from the Student Guide*)) All magnets have a magnetic field in which the force of the magnetic attraction can be felt. If we place iron filings near a magnet, the filings will line up in a pattern that shows the field produced by the magnet. In today's experiment, you are going to test to see if you can use electricity to yield the same effect.
- ☞ **Results –** The students will see that they are able to create a magnetic field with electricity using the electromagnet they built.
- ☞ **Explanation –** All electrical current creates a small magnetic field. When you wrap the wire in a coil around the nail, the magnetic field created is multiplied. The end result is a field similar to that found a bar magnet, with a north and south poles. The magnet created is a temporary magnet because as soon as the electricity (battery) is disconnected, the coil is demagnetized.
- ☞ **Take it Further –** Have the students test the strength of their electromagnet using the two tests from the previous week.

Discussion Questions

1. What does an electrical current always produce? (*An electrical current will always produce a magnetic field.*)
2. What are two benefits of using electromagnets over permanent magnets? (*Electromagnets can be turned off and on, whereas permanent magnets are always on. Plus, you can alter the attractive strength of an electromagnet by changing the current.*)
3. How does a magnetic levitation train work? (*A magnetic levitation train works by passing a current through electromagnets in the track and on the train. The magnetism that is produced by the current lifts the train up and pulls it along the track.*)
4. What happens when electricity flows through a coil of wires? (*When electricity flows through a coil of wires, the magnetic field combines to produce a magnetic field similar to a permanent bar magnet.*)
5. What are some items that we encounter in everyday life that use electromagnetism? (*Students' answers will vary, but they could include door latches, doorbells, metal detectors, car fuel gauges, or vending machines.*)

Want More

- **Magnets and Electromagnets –** Have the students do the "Magnets and Electromagnets" simulation from PhET. Here is the link to the on-line demonstration:
  - http://phet.colorado.edu/en/simulation/magnets-and-electromagnets
- **Electromagnetic Motor –** Have the students make their own electromagnetic motor with a battery, 2 safety pins, a rubber band, a bobbin, and copper wire. Watch the following YouTube video for directions:
  - https://www.youtube.com/watch?v=HHEdIZ282hE

Sketch Week 26

Magnetic Fields in an Electromagnet

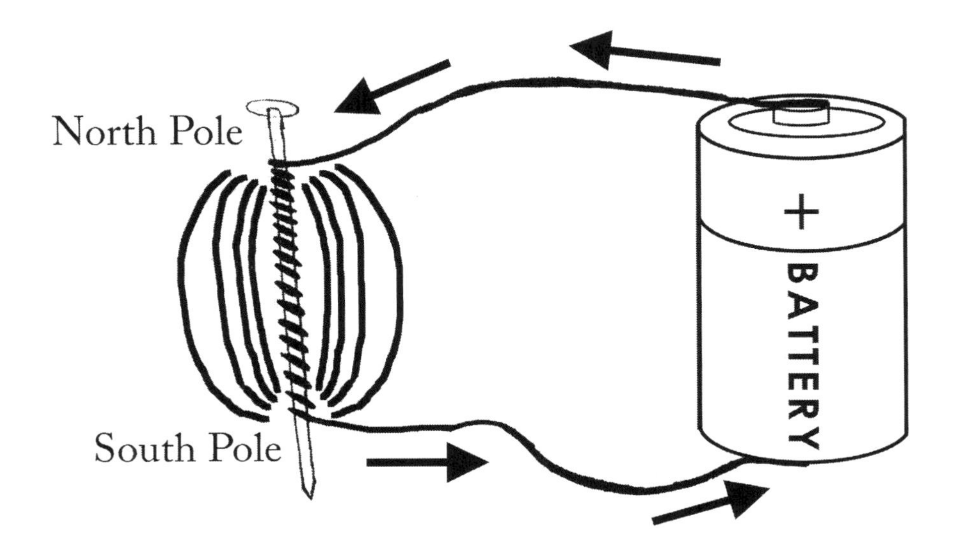

Student Assignment Sheet Week 27
Motors and Generators

Experiment: Can I raise a needle from a table without touching it?

Materials

- ✓ Straws
- ✓ Electrical tape
- ✓ 6 ft. of thin insulated wire
- ✓ AAA battery
- ✓ Sandpaper
- ✓ Needle
- ✓ *Robotics* book

Procedure

1. Read the introduction to the experiment on pg. 50 of *Robotics*.
2. Follow the directions on pp. 50-51 of *Robotics* to build your solenoid. Then, use the solenoid to attempt to raise a needle from a table without touching it.
3. Draw conclusions and complete the experiment sheet.

Vocabulary & Memory Work

- ☐ Vocabulary: power, solenoid, turbine
- ☐ Memory Work—There is no memory work this week.

Sketch: Fleming's Hand-rules

- Label the following – Left-hand rule for motors, thumb gives the wire's direction of motion, first finger shows the magnetic field's direction, second finger shows the current's direction, right-hand rule for generators, thumb give the direction of motion, first finger shows the magnetic field's direction, second finger shows the current's direction.

Writing

- Reading Assignment: *DK Encyclopedia of Science* pg. 158 (Electric Motors), pg. 159 (Generators)
- Additional Research Readings
  - 📖 Generators and Motors: *KSE* pp. 346-347
  - 📖 Electromagnetism (the motor effect): *UIDS* pg. 76

Dates

- 🕒 1821 – Michael Faraday discovers that electricity can produce rotary motion.

Schedules for Week 27

Two Days a Week

Day 1	Day 2
☐ Do the "Can I raise a needle from a table without touching it?" experiment, and then fill out the experiment sheet on SG pp. 196-197 ☐ Define power, solenoid, and turbine on SG pg. 157 ☐ Enter the dates onto the date sheets on SG pp. 8-13	☐ Read pp. 158-159 from *DK EOS,* and then discuss what was read ☐ Color and label the "Fleming's Hand-rules" sketch on SG pg. 195 ☐ Prepare an outline or narrative summary; write it on SG pp. 198-199

Supplies I Need for the Week

✓ Straws, Electrical tape, 6 ft. of thin insulated wire, AAA battery
✓ Sandpaper, Needle
✓ *Robotics* book

Things I Need to Prepare

Five Days a Week

Day 1	Day 2	Day 3	Day 4	Day 5
☐ Do the "Can I raise a needle from a table without touching it?" experiment, and then fill out the experiment sheet on SG pp. 196-197 ☐ Enter the dates onto the date sheets on SG pp. 8-13	☐ Read pp. 158-159 from *DK EOS,* and then discuss what was read ☐ Write an outline on SG pg. 198	☐ Define power, solenoid, and turbine on SG pg. 157 ☐ Color and label the "Fleming's Hand-rules" sketch on SG pg. 195	☐ Read one or all of the additional reading assignments ☐ Write a report on what you learned on SG pg. 199	☐ Complete one of the Want More Activities listed **OR** ☐ Study a scientist from the field of Physics

Supplies I Need for the Week

✓ Straws, Electrical tape, 6 ft. of thin insulated wire, AAA battery
✓ Sandpaper, Needle
✓ *Robotics* book

Things I Need to Prepare

Additional Information Week 27

Experiment Information

- **Introduction –** (*from the Student Guide*) Read the introduction about solenoids found on pg. 50 of *Robotics*.
- **Results –** The students should see that they were able to suck the needle into the straw without using their hands.
- **Explanation –** The students created an electromagnet around the end of the straw. When electricity flowed through the wires, the copper was temporarily magnetized and the needle was drawn in by the attractive force of the magnetic field that was created.
- **Take it Further –** Have the students repeat the experiment using a nail and strong magnet, as suggested in the *Robotics* book.

Discussion Questions

Motors, pg. 158

1. What does an electric motor do? (*An electric motor turns electricity into movement.*)
2. How does an electric motor work? (*An electric motor takes advantage of the fact that a current-carrying wire produces a magnetic field. The magnetic field produced in a wire coil interacts with another permanent magnetic field, causing the coil to turn. To keep the movement going, the current is reversed every half turn.*)
3. Why are electric motors good sources of power? (*Electric motors are clean, quiet, and versatile.*)

Generators, pg. 159

1. How do generators differ from motors? (*Generators work opposite from motors, in that they turn movement into electricity.*)
2. What is electromagnetic induction? (*Electromagnetic induction is when electricity is produced in a wire as it moves in a magnetic field.*)

Want More

- **Reverse Engineer –** Have the students reverse engineer a small motor, like one in a discarded remote control car or a fan from an old computer. As the students take apart the motor, have them look for the different parts they have studied. You can use the information and printable worksheets from the following post at the Homeschool Scientist as a guide:
  - http://thehomeschoolscientist.com/reverse-engineering-printable-worksheets/

Sketch Week 27

Fleming's Hand-rules

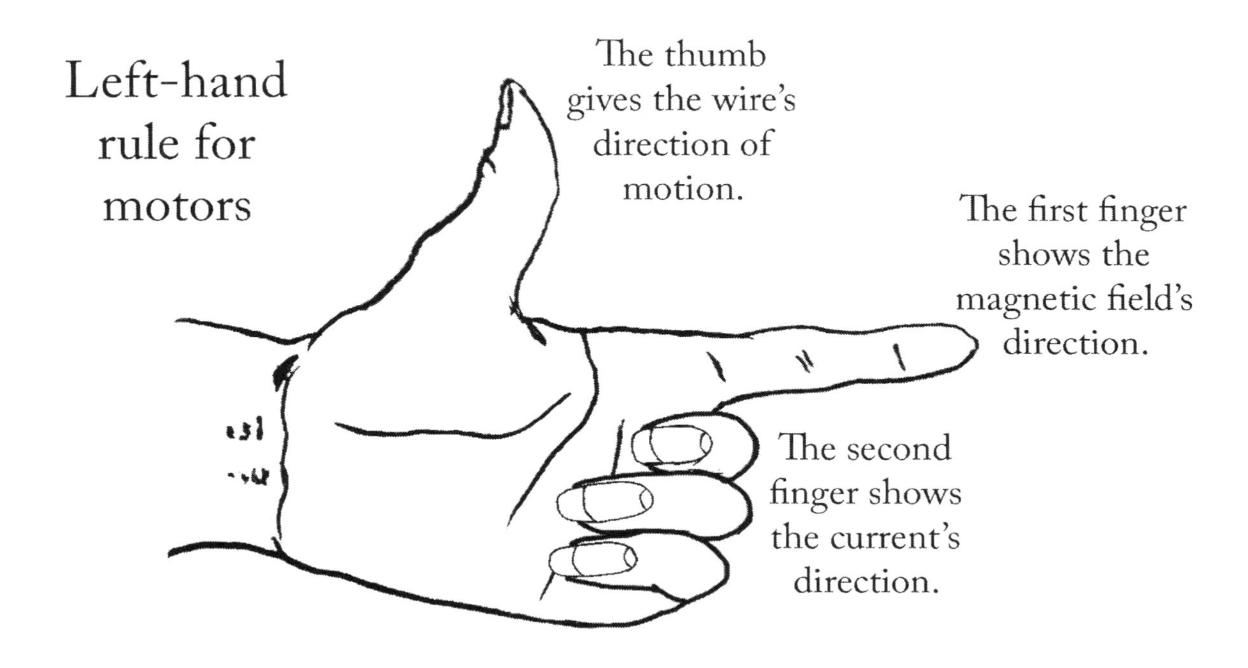

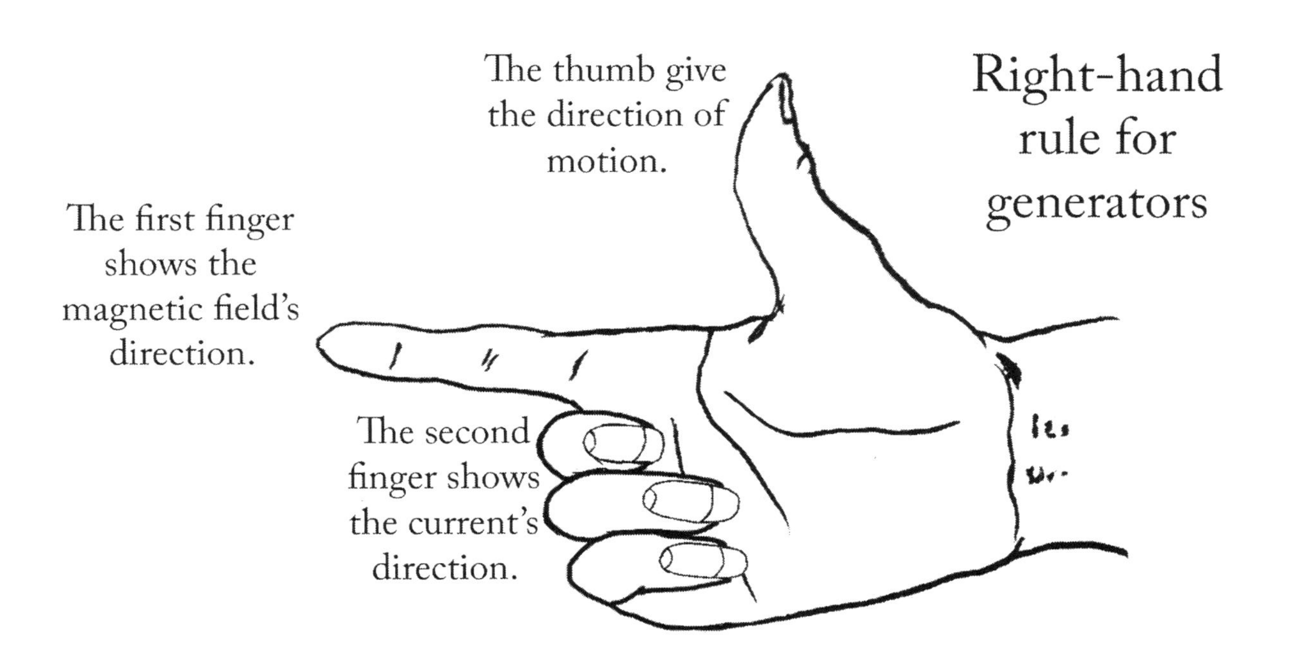

Unit 6 Electricity and Magnetism
Unit Test Answers

Vocabulary Matching

1. M
2. L
3. A
4. F
5. I
6. B
7. G
8. J
9. P
10. N
11. C
12. O
13. H
14. D
15. E
16. R
17. K
18. Q

True or False

1. True
2. False (*When you charge an object through friction, the electrons from one object are rubbed off onto another object, creating a static charge.*)
3. True
4. True
5. False (*All batteries store chemical energy and change it into electrical energy.*)
6. True
7. True
8. False (*The greater the resistance in a circuit, the less the current flows.*)
9. False (*The north pole of one magnet is always attracted to the south pole of another magnet.*)
10. True
11. False (A*n electrical current will always produce a magnetic field.*)
12. True
13. False (*An electric motor turns electricity into movement.*)
14. False (*A generator turns movement into electricity.*)

Short Answer

1. Attraction is when two objects with opposite electrical charges move towards one another. Repulsion is when two objects with the same electrical charge move away from one another.
2. In electroplating, a current runs through the solution and the material, such as circuit boards. The material acts as an anode, which attracts the positive metal ions, causing the metal ions to attach to the material, thereby creating a thin layer of the metal on the material.
3. The three basic components of a cell are the positive electrode, the negative electrode, and an electrolyte chemical.
4. When resistors are connected in series, the overall resistance in a circuit increases. When resistors are connected in parallel, the overall resistance in a circuit decreases.
5. The Earth's magnetic field is cause by the iron core at the center of the Earth. The Earth's magnetic field allows us to use a compass to navigate. It also causes an aurora to be produced, known as the northern lights.
6. Electromagnets can be turned off and on, whereas permanent magnets are always on. Plus, you can alter the attractive strength of an electromagnet by changing the current.

7. An electric motor takes advantage of the fact that a current-carrying wire produces a magnetic field. The magnetic field produced in a wire coil interacts with another permanent magnetic field, causing the coil to turn. To keep the movement going, the current is reversed every half turn.
8. Direct Current is when the electrical charge only flows in one direction. Alternating Current is when the electrical charge regularly reverses the direction of its flow.
9. Like charges repel each other and opposite charges attract each other.

Unit 6 Electricity and Magnetism
Unit Test

Vocabulary Matching

1. Electrical charge ____
2. Induction ____
3. Conductor ____
4. Insulator ____
5. Semiconductor ____
6. Anode ____
7. Cathode ____
8. Diode ____
9. Parallel circuit ____
10. Resistance ____
11. Series circuit ____
12. Electrolyte ____
13. Magnetic field ____
14. Magnetic pole ____
15. Electromagnet ____
16. Electromagnetic force ____
17. Solenoid ____
18. Turbine ____

A. A material through which electrical charge can easily flow.

B. A positively charged diode.

C. An electrical circuit in which current passes through components that are one after another.

D. One of the two ends of a magnet where the force of attraction or repulsion is the strongest.

E. A magnet that can be switched off and on with electric current.

F. A material through which electrical charge cannot easily flow.

G. A negatively charged diode.

H. The area around a magnet in which the magnetic force can be felt.

I. A material that can act as a conductor or insulator based on its temperature.

J. An electrical component that allows electrical current to flow in and out in a single direction.

K. A coil of wire that behaves like a magnet when electric current passes through it.

L. The transfer of electrical charge without contact between the materials.

M. A charge produced by an excess or shortage of electrons.

N. The ability of a material to resist the flow of electrical current.

O. A substance that can conduct electrical current.

P. An electrical circuit in which current can past though more than one path.

Q. A machine with shafts and blades that are turned by the force of wind or steam; this movement generates energy that can be turned into electricity.

R. The force produced when an electrical current flows through a wire; it forms a magnetic field.

True or False

1. ____________________ Lodestone is a mineral that is naturally magnetized.

2. ____________________ When you charge an object through friction, the protons from one object are rubbed off onto another object, creating a static charge.

3. ____________________ Electric current is the flow of electricity from one place to another.

4. ____________________ To "dope" a semiconductor means to make it impure.

5. ____________________ All batteries store kinetic energy and change it into potential energy.

6. ____________________ A larger cell battery produces the same electromotive force, but it last longer than a smaller cell.

7. ____________________ A circuit is the path that electricity takes when it flows.

8. ____________________ The greater the resistance in a circuit, the more the current flows.

9. ____________________ The north pole of one magnet is always attracted to the north pole of another magnet.

10. ____________________ A piece of steel will magnetize when you stroke it with a magnet, causing all the domains in the bar to align in the same direction.

11. ____________________ An electrical current is not able to produce a magnetic field.

12. ____________________ When electricity flows through a coil of wires, the magnetic field combines to produce a magnetic field similar to a permanent bar magnet.

13. ____________________ An electric motor turns movement into electricity.

14. ____________________ A generator turns electricity into movement.

Short Answer

1. What are attraction and repulsion?

2. How does electroplating work?

3. What are the three basic components of a cell?

4. How is resistance affected when the resistors are connected in series? In parallel?

5. What causes the Earth's magnetic field? How does the magnetic field affect life on the surface of the Earth?

6. What are two benefits of using electromagnets over permanent magnets?

7. How does an electric motor work?

8. What is the difference between direct and alternating current?

9. Write the law of electrostatics.

Physics Unit 7

Engineering and Robotics

Unit 7 Engineering and Robotics
Overview of Study

Sequence of Study

Week 28: Engineering
Week 29: Bridges
Week 30: Tunnels
Week 31: Robotics
Week 32: Actuators and Effectors
Week 33: Sensors and Controllers

Materials by Week

Week	Materials
28	Paper, Tape, Books, Can or glass
29	Craft sticks, Wood glue, Books, Binder clips
30	Salt dough (at least 3 to 4 cups), Cardboard square, Spoon, Craft sticks, Pipe cleaners, Aluminum foil, Toy car, Books or other heavy objects, Water
31	1.5-volt DC motor, 1 ft. insulated wire, Electrical tape, Cup, Foam tape, 2 AAA batteries, Rubber band, Cork, Marker, Cardboard, Paper
32	Pencil, 1.5-volt DC motor, Small Solar Panel, Electrical tape, Scissors, CD, Glue, Tape, Clear dome from a drink cup
33	1 LED light bulb with two metal legs, 1 3-volt Watch battery, 2 Index cards, Scissors, Marker, Yarn, Gel glue, Toothpick, Tissue

Vocabulary for the Unit

1. **Engineer** – A person who uses science and math to design, build, and maintain engines, machines, bridges, tunnels, and other public works structures.
2. **Load** – An applied force or weight.
3. **Beam** – A horizontal, weight-bearing, rigid structure that carries a load.
4. **Bridge** – A man-made structure that has been built to span rivers, canyons, roads, railways, and more.
5. **Truss** – A framework of beams of bars that can support structures, like bridges.
6. **Arch** – A curved weight-bearing structure that is in the shape of an upside-down U.
7. **Tunnel** – A passageway that goes under or through a natural or man-made obstacle, such as rivers, mountains, building, roads, and more.
8. **Automaton** – A machine that can move by itself.
9. **Robot** – A machine that is able to sense, think, and act on its own.
10. **Actuator** – A piece of equipment that makes a robot move.
11. **Capacitor** – An electrical component that stores electrical charge and releases the charge when it is needed.
12. **Effector** – A device that lets a robot affect things in the world around it, such as grippers, tools, and laser beams.

13. **Controller –** A computer or switch that can react to the information gathered by sensors.
14. **Photoresistor –** A light sensor that is able to change the resistance in an electrical current depending upon the amount of light it receives.
15. **Sensor –** A device in robotics that takes in information from the outside world.

Memory Work for the Unit

The Engineering Design Process

1. Identify the Need or Problem
2. Research and Define Requirements
3. Brainstorm for Solutions
4. Design a Plan
5. Build a Prototype
6. Test and Evaluate the Prototype
7. Communicate the Results
8. Redesign if Needed

Notes

Student Assignment Sheet Week 28
Engineering

Experiment: Which shape is stronger?

Materials

- ✓ Paper
- ✓ Tape
- ✓ Books
- ✓ Can or glass

Procedure

1. Read the introduction to the experiment on pg. 22 of *Bridges and Tunnels*.
2. Follow the directions on pg. 22 of *Bridges and Tunnels* to make your shapes. Then, test which one of the shapes you made is stronger.
3. Draw conclusions and complete the experiment sheet.

Vocabulary & Memory Work

- ☐ Vocabulary: engineer, load
- ☐ Memory Work—This week, work on memorizing the engineering design process.
  - **The Engineering Design Process**
    1. Identify the Need or Problem
    2. Research and Define Requirements
    3. Brainstorm for Solutions
    4. Design a Plan
    5. Build a Prototype
    6. Test and Evaluate the Prototype
    7. Communicate the Results
    8. Redesign if Needed

Sketch: 5 Main Branches of Engineering

- Use the information found in the chart on pg. 9 of *Bridges and Tunnels* to fill in the chart.

Writing

- Reading Assignment: *Bridges and Tunnels* pp. 1-6 Lifelines, pp. 7-17 Engineering and Thinking Big
- Additional Research Readings
  - Brick, Stone, and Concrete: *KSE* pp. 220-221

Dates

- 520 BC – Greek architect, Eupalinos, becomes the first engineer.
- 27 BC-393 AD – The Roman Empire constructs many arch bridges, some of which still stand today.

Schedules for Week 28

Two Days a Week

Day 1	Day 2
☐ Do the "Which shape is stronger?" experiment, and then fill out the experiment sheet on SG pp. 206-207 ☐ Define engineer and load on SG pg. 202 ☐ Enter the dates onto the date sheets on SG pp. 8-13	☐ Read the Introduction and Chapter 1 of *Bridges and Tunnels,* and then discuss what was read ☐ Fill out the "5 Main Branches of Engineering" chart on SG pg. 205 ☐ Prepare an outline or narrative summary, write it on SG pp. 208-209
Supplies I Need for the Week ✓ Paper, Tape ✓ Books ✓ Can or glass	
Things I Need to Prepare	

Five Days a Week

Day 1	Day 2	Day 3	Day 4	Day 5
☐ Do the "Which shape is stronger?" experiment, and then fill out the experiment sheet on SG pp. 206-207 ☐ Enter the dates onto the date sheets on SG pp. 8-13	☐ Read the Introduction and Chapter 1 of *Bridges and Tunnels,* and then discuss what was read ☐ Write an outline on SG pg. 208	☐ Define engineer and load on SG pg. 202 ☐ Fill out the "5 Main Branches of Engineering" chart on SG pg. 205	☐ Read one or all of the additional reading assignments ☐ Write a report on what you learned on SG pg. 209	☐ Complete one of the Want More Activities listed **OR** ☐ Study a scientist from the field of Physics
Supplies I Need for the Week ✓ Paper, Tape ✓ Books ✓ Can or glass				
Things I Need to Prepare				

Additional Information Week 28

Experiment Information

☞ **Introduction** – (*from the Student Guide*) Read the introduction about engineers and shapes found on pg. 22 of *Bridges and Tunnels*.

☞ **Results** – The students' results will vary based on which shapes they choose, the number of shapes they construct, and how they arrange the shapes under the book. Generally, columns are extremely strong and the more columns you have supporting the books, the more weight the structure can hold.

☞ **Explanation** – Structures fail for three main reasons: buckling, compression, or shear. In buckling, the support bends near the middle due to the weight. In compression, the support compresses under the weight. In shear, the support breaks off under the weight. The paper supports built by the students are most likely to experience buckling or compression under the weight of the books.

☞ **Take it Further** – Have the students do the building materials assessment suggested on pp. 18-19 of the *Bridges and Tunnels* book.

Discussion Questions

Lifelines, pp. 1-6

1. What do bridges and tunnels do? (*Bridges and tunnels connect people and places.*)
2. What is trial-and-error? (*Trial-and-error is a method of engineering that tries one design and then another until the engineer finds one that works.*)

Engineering and Thinking Big, pp. 7-17

1. What are the different branches of engineering? (*The main branches of engineering are chemical, civil, computer, electrical, and mechanical.*)
2. What is tension? (*Tension is a force that pulls or stretches a material outward.*) Compression? (*Compression is a force that squeezes or presses a material inward.*)
3. How do tension and compression affect a structure? (*Tension and compression are the two main forces that keep a structure standing.*)
4. What is torsion? (*Torsion is a force that twists and turns a material.*) Shear? (*Shear is a force that slips parts of a material in opposite directions.*)
5. What natural phenomena can produce torsion and shear within a structure? (*Wind and earthquakes can produce torsion and shear in a structure, which can result in damage.*)

Want More

❐ **Egg Bungee Drop** – Have the students do the egg bungee drop activity suggested on pp. 20-21 of *Bridges and Tunnels*.

5 Main Branches of Engineering

Chemical	Chemical engineers use chemicals and raw materials and turn them into food and energy that people can use.
Civil	Civil engineers design and build bridges, tunnels, dams, roads, and more for civil use.
Computer	Computer engineers develop technology used in computers, such as software, hardware, operating systems, and networks.
Electrical	Electrical engineers design electronic products and electrical systems.
Mechanical	Mechanical engineers develop mechanical systems, like engines, tools, and machines.

Student Assignment Sheet Week 29
Bridges

Experiment: How strong is a craft stick beam bridge?

Materials
- ✓ Craft sticks
- ✓ Wood glue
- ✓ Books
- ✓ Binder clips

Procedure
1. Read the introduction to the experiment on pg. 58 of *Bridges and Tunnels*.
2. Follow the directions on pg. 58-59 of *Bridges and Tunnels* to build your craft stick beam bridge. Then, test how strong the bridge is.
3. Draw conclusions and complete the experiment sheet.

Vocabulary & Memory Work
- ☐ Vocabulary: beam, bridge, truss
- ☐ Memory Work—This week, continue to work on memorizing the engineering design process.

Sketch: 3 Types of Bridges
- Label the following – beam bridge, arch bridge, suspension bridge

Writing
- Reading Assignment: *Bridges and Tunnels* pp. 23-31 Building Big: The Physics of Bridges, pp. 39-56 Amazing Bridges
- Additional Research Readings
  - Construction: *KSE* pp. 223-223
  - When Bridges Collapse: *Bridges and Tunnels* Chapter 4

Dates
- 62 BC – The Ponte Fabrico Bridge, a stone double-arch bridge, is built by the Roman Empire.
- 1848 – One of the earliest wire suspension bridges is designed by Charles Ellet and built to span the Niagara River near Niagara Falls.
- 1949 – The world's first geodesic dome is built by Richard Fuller.

Schedules for Week 29

Two Days a Week

Day 1	Day 2
☐ Do the "How strong is a craft stick beam bridge?" experiment, and then fill out the experiment sheet on SG pp. 212-213 ☐ Define beam, bridge, and truss on SG pg. 202 ☐ Enter the dates onto the date sheets on SG pp. 8-13	☐ Read the Chapters 2 and 3 of *Bridges and Tunnels,* and then discuss what was read ☐ Color and label the "3 Types of Bridges" sketch on SG pg. 211 ☐ Prepare an outline or narrative summary; write it on SG pp. 214-215
Supplies I Need for the Week ✓ Craft sticks, Wood glue ✓ Books, Binder clips	
Things I Need to Prepare	

Five Days a Week

Day 1	Day 2	Day 3	Day 4	Day 5
☐ Do the "How strong is a craft stick beam bridge?" experiment, and then fill out the experiment sheet on SG pp. 212-213 ☐ Enter the dates onto the date sheets on SG pp. 8-13	☐ Read the Chapters 2 and 3 of *Bridges and Tunnels,* and then discuss what was read ☐ Write an outline on SG pg. 214	☐ Define beam, bridge, and truss on SG pg. 202 ☐ Color and label the "3 Types of Bridges" sketch on SG pg. 211	☐ Read one or all of the additional reading assignments ☐ Write a report on what you learned on SG pg. 215	☐ Complete one of the Want More Activities listed **OR** ☐ Study a scientist from the field of Physics
Supplies I Need for the Week ✓ Craft sticks, Wood glue ✓ Books, Binder clips				
Things I Need to Prepare				

Additional Information Week 29

Experiment Information

- **Introduction –** (*from the Student Guide*) Read the introduction about beam bridges found on pg. 58 of *Bridges and Tunnels*.
- **Results –** The students' results will vary based on how they construct their bridges.
- **Explanation –** Beam bridges have two types of supports, both of which are there to prevent buckling. Columns support the beams and decking from underneath. Trussing provides additional rigidity to the beam from above. The more support and rigidity the beam bridge has, the more weight it can hold.
- **Take it Further –** Have the students modify their designs to hold even more weight than before.

Discussion Questions

Building Big: The Physics of Bridges, pp. 23-31

1. What are the three major bridge designs? (*The three major bridge designs are beam, arch, and suspension.*)
2. How is a beam bridge constructed and when is it useful? (*The beam bridge has a long, rigid, horizontal beam and two vertical abutments that support the beam at either end. Beam bridges are simple and cheap to build, but they cannot span long distances.*)
3. How is an arch bridge constructed and when is it useful? (*The arch bridge has abutments like the beam bridge, but it uses an arch or a series of arches that transfer the load to the abutments. These bridges can span longer distances than beam bridges can.*)
4. How is a suspension bridge constructed and when is it useful? (*The suspension bridge has anchorage points at either end where cables are attached. These cables are strung across and down to hold the roadway underneath. These bridges are extremely sturdy and can span quite long distances.*)

Amazing Bridges, pp. 39-56

1. What are caissons and cofferdams? (*Caissons and cofferdams are watertight structures that builders use while digging out and pouring the supports for a bridge that spans a waterway.*)
2. What does the keystone do in an arch? (*The keystone locks the stones in the arch in place.*)
3. Tell the story of how the Brooklyn Bridge or Golden Gate Bridge was built. (*Answers will vary.*)

Want More

- **Geodesic Dome –** Have the students build a geodesic dome out of marshmallows using the directions found on pg. 32 of *Bridges and Tunnels*.
- **Materials Testing –** Have the students test how different construction materials hold up under the stress of compression. Use the activity suggested on pp. 34-35 of *Bridges and Tunnels*.
- **Penny Bridge –** Have the students build a bridge out of pennies using the directions found on pg. 69 of *Bridges and Tunnels*.

Sketch Week 29

3 Types of Bridges

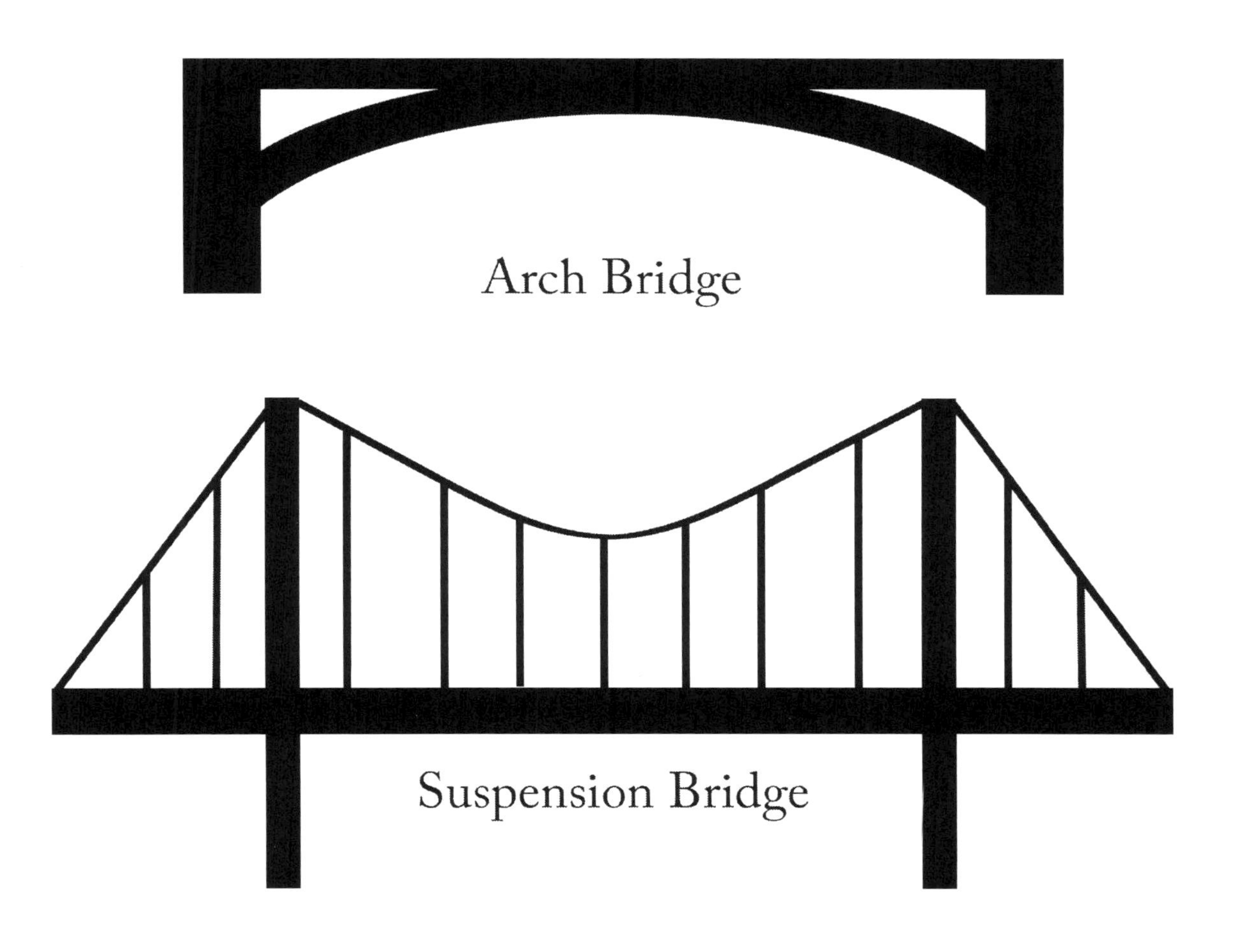

Student Assignment Sheet Week 30
Tunnels

Experiment: Tunnel Building Project

Materials

- ✓ Salt dough (at least 3 to 4 cups, recipe for a homemade version of this can be found in the Appendix on pg. 269)
- ✓ Spoon
- ✓ Craft sticks
- ✓ Pipe cleaners
- ✓ Aluminum foil
- ✓ Cardboard square
- ✓ Toy car
- ✓ Books or other heavy objects
- ✓ Water

Procedure

1. Read the introduction to the experiment.
2. Set the salt dough on the cardboard square and shape the dough into a mountain.
3. Then, use the spoon to dig a tunnel at the base of the mountain large enough for the toy car to drive through. Add supports, pipe cleaners, and craft sticks, and a foil lining for the tunnel as you dig.
4. Once the tunnel is complete, set the mountain aside and let it dry for several days.
5. After the mountain is dry, head outside and place the toy car in the tunnel. Then, slowly pour water over your mountain and check to see if any water seeps into your tunnel.
6. Next, use the books or other heavy objects to test the whether your tunnel will collapse.
7. Draw conclusions and complete the experiment sheet.

Vocabulary & Memory Work

- ☐ Vocabulary: arch, tunnel
- ☐ Memory Work—This week, continue to work on memorizing the engineering design process.

Sketch: Forces in a Tunnel

- Draw and label the forces that are present in a tunnel – compression forces from the weight above, pushing forces from the solid rock weight below

Writing

- Reading Assignment: *Bridges and Tunnels* pp. 74-84 Building Big: The Physics of Tunnels
- Additional Research Readings
  - Amazing Tunnels: *Bridges and Tunnels* Chapter 6
  - Tunnel Disasters: *Bridges and Tunnels* Chapter 7

Dates

- 1869-1883 – The Brooklyn Bridge, a steel-cable suspension bridge, is built.
- 1899 – Alfred Nobel invents dynamite, which changes tunnel and bridge building forever.

Schedules for Week 30

Two Days a Week

Day 1	Day 2
☐ Do the "Tunnel Building Project," and then fill out the experiment sheet on SG pp. 218-219 ☐ Define arch and tunnel on SG pg. 202 ☐ Enter the dates onto the date sheets on SG pp. 8-13	☐ Read the Chapter 5 of *Bridges and Tunnels,* and then discuss what was read ☐ Color and label the "Forces in a Tunnel" sketch on SG pg. 217 ☐ Prepare an outline or narrative summary; write it on SG pp. 220-221

Supplies I Need for the Week

- ✓ Salt dough (at least 3 to 4 cups), Cardboard square
- ✓ Spoon, Craft sticks, Pipe cleaners, Aluminum foil,
- ✓ Toy car, Books or other heavy objects, Water

Things I Need to Prepare

Five Days a Week

Day 1	Day 2	Day 3	Day 4	Day 5
☐ Do the "Tunnel Building Project," and then fill out the experiment sheet on SG pp. 218-219 ☐ Enter the dates onto the date sheets on SG pp. 8-13	☐ Read the Chapter 5 of *Bridges and Tunnels,* and then discuss what was read ☐ Write an outline on SG pg. 220	☐ Define arch and tunnel on SG pg. 202 ☐ Color and label the "Forces in a Tunnel" sketch on SG pg. 217	☐ Read one or all of the additional reading assignments ☐ Write a report on what you learned on SG pg. 221	☐ Complete one of the Want More Activities listed **OR** ☐ Study a scientist from the field of Physics

Supplies I Need for the Week

- ✓ Salt dough (at least 3 to 4 cups), Cardboard square
- ✓ Spoon, Craft sticks, Pipe cleaners, Aluminum foil,
- ✓ Toy car, Books or other heavy objects, Water

Things I Need to Prepare

Additional Information Week 30

Experiment Information

- ☞ **Introduction –** (*from the Student Guide*) Tunnels are built to get through mountains and other barriers. Engineers have to design supports that prevent the tunnel from collapsing and keep water from filling the tunnel. In today's experiment, you will be engineering your own tunnel through a salt dough mountain.
- ☞ **Results and Explanation –** The students' results will vary based on their designs. Be sure to discuss their results and talk about how they could improve on their designs.
- ☞ **Take it Further –** Have the students design another salt dough tunnel that goes under water. This time, use a deep plastic bin and shape the dough to have two ends with land and a rounded basin in the middle for water. Be sure to leave enough room at the bottom for the tunnel to go through.

Discussion Questions

1. What are two ways that natural tunnels can be formed? (*Natural tunnels can be dug out by animals or formed by lava.*)
2. What are some ways that humans have used tunnels? (*Humans have used tunnels to irrigate crops, bring in fresh water, remove sewage, get through mountains, and go under water or roadways.*)
3. What are the three stages of man-made tunnel building? Explain what happens in each. (*The three stages of man-made tunnel building are excavation, support, and lining. In the excavation stage, the engineers bore, or hollow out, the passageway. In the support stage, the engineers build supports that prevent collapses and water seepage. In the lining stage, the engineers line the tunnel to provide a permanent support for the tunnel.*)
4. What are the three tunnel types? (*The three tunnel types are soft-ground, hard-rock, and underwater.*)

Want More

- **Tunnel building and the Bends –** Have the students research about what the bends is and how it affects a human. Then, have them do the activity suggested on pg. 85 of *Bridges and Tunnels*.
- **Aquifer –** Have the students learn more about aquifers by doing the activity suggested on pp. 104-105 of *Bridges and Tunnels*.

Sketch Week 30

Forces in a Tunnel

Compression forces from the weight above.

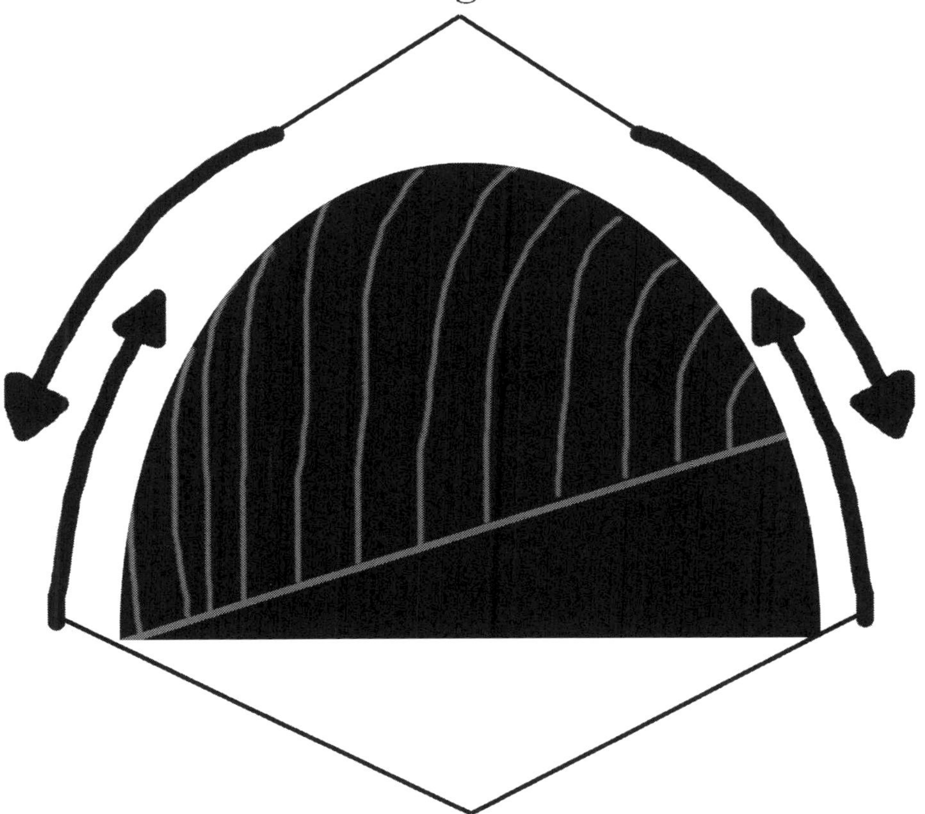

Pushing forces from the solid rock weight below.

Student Assignment Sheet Week 31
Robotics

Experiment: Vibrorobot

Materials
- ✓ 1.5-volt DC motor
- ✓ 1 ft. insulated wire
- ✓ Electrical tape
- ✓ Cup or Jar
- ✓ Foam tape
- ✓ 2 AAA batteries
- ✓ Rubber band
- ✓ Cork
- ✓ Cardboard
- ✓ 3 Pens
- ✓ Paper

Procedure
1. Read the introduction to the experiment on pg. 24 of *Robotics*.
2. Follow the directions on pg. 24-26 of *Robotics* to build a vibrating robot. Then, test how the robot works as it draws a picture for you.
3. Draw conclusions and complete the experiment sheet.

Vocabulary & Memory Work
- ❒ Vocabulary: automata, robot
- ❒ Memory Work—There is no memory work for this week.

Sketch: Sense – Think – Act Cycle
- Label the following – the robot takes in information, the robot uses the information to choose the next step, the robot does something.

Writing
- Reading Assignment: *Robotics* pp. 11-23 Development of Robotics, pp. 27-33 Housing: Robot Bodies
- Additional Research Readings
  - 🕮 Introduction: *Robotics* pp. 1-7

Dates
- 200 BC – The Chinese create mechanical musicians to play music for the emperor.
- 1555 – Gianello Torriano, an Italian clock maker, makes a model of a wind-up lady that can walk in a circle and strum a lute.
- 1822 – Charles Babbage designs a mechanical calculator that uses punch cards.
- 1959 – The Massachusetts Institute of Technology (MIT) opens the first lab to study artificial intelligence.

Schedules for Week 31

Two Days a Week

Day 1	Day 2
☐ Build the "Vibrorobot," and then fill out the experiment sheet on SG pp. 224-225 ☐ Define automata and robots on SG pp. 202-203 ☐ Enter the dates onto the date sheets on SG pp. 8-13	☐ Read the Chapter 1 and 2 of *Robotics,* and then discuss what was read ☐ Color and label the "Sense - Think - Act Cycle" sketch on SG pg. 223 ☐ Prepare an outline or narrative summary; write it on SG pp. 226-227
Supplies I Need for the Week ✓ 1.5-volt DC motor, 1 ft. insulated wire, Electrical tape ✓ Cup or Jar, Foam tape, 2 AAA batteries, Rubber band, ✓ Cork, Cardboard, 3 Pens, Paper	
Things I Need to Prepare	

Five Days a Week

Day 1	Day 2	Day 3	Day 4	Day 5
☐ Build the "Vibrorobot," and then fill out the experiment sheet on SG pp. 224-225 ☐ Enter the dates onto the date sheets on SG pp. 8-13	☐ Read the Chapter 1 and 2 of *Robotics,* and then discuss what was read ☐ Write an outline on SG pg. 226	☐ Define automata and robots on SG pp. 202-203 ☐ Color and label the "Sense - Think - Act Cycle" sketch on SG pg. 223	☐ Read one or all of the additional reading assignments ☐ Write a report on what you learned on SG pg. 227	☐ Complete one of the Want More Activities listed **OR** ☐ Study a scientist from the field of Physics
Supplies I Need for the Week ✓ 1.5-volt DC motor, 1 ft. insulated wire, Electrical tape ✓ Cup or Jar, Foam tape, 2 AAA batteries, Rubber band, ✓ Cork, Cardboard, 3 Pens, Paper				
Things I Need to Prepare				

Additional Information Week 31

Experiment Information

- ☞ **Introduction –** (*from the Student Guide*) Read the introduction on vibrorobots found on pg. 24 of *Robotics*.
- ☞ **Results and Explanation –** The students' results will vary based on their designs. Be sure to discuss their results and talk about how they could improve on their designs.
- ☞ **Troubleshooting –** Make sure the motor is in the center of the cup or jar so that the robot will vibrate at its best.
- ☞ **Take it Further –** Have the students work on improving their designs using the suggestions found on pg. 26 of *Robotics*.

Discussion Questions

Development of Robots, pp. 11-23

1. What kinds of automata existed before modern-day robots? (*Before modern-day robots, people built automata like mechanical musicians, knights, and ladies that worked using wind, water, and gears.*)
2. What do people use robots for? (*Answers will vary, but should include examples of people using robots in homes, toys, art, medicine, industry, military, and space.*)

Housing: Robot Bodies, pp. 27-33

1. What are nanobots? (*Nanobots are robots that are too small to see without a microscope*.)
2. What is a robotic swarm? (*A robotic swarm is a group of small robots that work together*.)
3. What is a biomimetic robot? (*A biomimetic robot is a robot that copies a living thing in some way*.)

Want More

- **Robot or Not –** Have the students determine whether an object is a robot by doing the activity suggested on pp. 8-10 of *Robotics*.
- **Robot Skin –** Have the students make their own robot skins by doing the activity suggested on pp. 34-35 of *Robotics*.

Sense - Think - Act Cycle

Sense

The robot takes in information.

Think

The robot uses the information to choose the next step.

Act

The robot does something.

Student Assignment Sheet Week 32
Actuators and Effectors

Experiment: Wobblebot

Materials

- ✓ Pencil
- ✓ 1.5-volt DC motor
- ✓ Small Solar Panel
- ✓ Electrical tape
- ✓ Scissors
- ✓ CD
- ✓ Glue
- ✓ Tape
- ✓ Clear dome from a drink cup

Procedure

1. Read the introduction to the experiment on pg. 48 of *Robotics*.
2. Follow the directions on pg. 48-49 of *Robotics* to build a solar wobblebot. Then, test how the robot works.
3. Draw conclusions and complete the experiment sheet.

Vocabulary & Memory Work

- ☐ Vocabulary: actuator, capacitor, effector
- ☐ Memory Work—There is no memory work for this week.

Sketch: Anatomy of a Solar Cell

- Label the following – solar cell, junction, negative panel, positive panel, the sun knocks electrons off atoms and sets them in motion, the electrons are used to power a device.

Writing

- Reading Assignment: *Robotics* pp. 38-47 Actuators: Making Robots Move, pp. 55-59 Effectors: How Robots Do Things
- Additional Research Readings
  - 🕮 Robotics: *KSE* pp. 236-237
  - 🕮 Robots: *DK EOS* pg. 176

Dates

- 1989 – Mark Tilden coins the term "BEAM robot" to refer to a simple solar-powered, life-like robot.
- 2000 – The Honda car company develops ASIMO, which revolutionized the way robots get around.

Schedules for Week 32

Two Days a Week

Day 1	Day 2
☐ Build the "Wobblebot," and then fill out the experiment sheet on SG pp. 230-231 ☐ Define actuator, capacitor, and effector on SG pg. 203 ☐ Enter the dates onto the date sheets on SG pp. 8-13	☐ Read the Chapter 3 and 4 of *Robotics,* and then discuss what was read ☐ Color and label the "Anatomy of a Solar Cell" sketch on SG pg. 229 ☐ Prepare an outline or narrative summary; write it on SG pp. 232-233
Supplies I Need for the Week ✓ Pencil, 1.5-volt DC motor, Small Solar Panel ✓ Electrical tape, Scissors, CD ✓ Glue, Tape, Clear dome from a drink cup	
Things I Need to Prepare	

Five Days a Week

Day 1	Day 2	Day 3	Day 4	Day 5
☐ Build the "Wobblebot," and then fill out the experiment sheet on SG pp. 230-231 ☐ Enter the dates onto the date sheets on SG pp. 8-13	☐ Read the Chapter 3 and 4 of *Robotics,* and then discuss what was read ☐ Write an outline on SG pg. 233	☐ Define actuator, capacitor, and effector on SG pg. 203 ☐ Color and label the "Anatomy of a Solar Cell" sketch on SG pg. 229	☐ Read one or all of the additional reading assignments ☐ Write a report on what you learned on SG pg. 234	☐ Complete one of the Want More Activities listed **OR** ☐ Study a scientist from the field of Physics
Supplies I Need for the Week ✓ Pencil, 1.5-volt DC motor, Small Solar Panel ✓ Electrical tape, Scissors, CD ✓ Glue, Tape, Clear dome from a drink cup				
Things I Need to Prepare				

Additional Information Week 32

Experiment Information

- ☞ **Introduction –** (*from the Student Guide*) Read the introduction on BEAM robots found on pg. 48 of *Robotics*.
- ☞ **Results and Explanation –** The students' results will vary based on their designs. Be sure to discuss their results and talk about how they could improve on their designs.
- ☞ **Troubleshooting –** Make sure the robot is in full sun or it will not work.
- ☞ **Take it Further –** Have the students work on improving their designs using the suggestions found on pg. 49 of *Robotics*.

Discussion Questions

Actuators: Making Robots Move, pp. 38-47

1. What does BEAM stand for? (*The "B" in BEAM stands for biology or biomimetic. The "E" in BEAM stands for electronics. The "A" in BEAM stands for aesthetics or artistic. The "M" in BEAM stands for mechanics.*)
2. How does the addition of a capacitor help a BEAM robot? (*The addition of a capacitor in a BEAM robot allows the machine to store power until there is enough to make it move.*)
3. What are two ways, other than a battery, to power a robot? (*In addition to battery power, you can use solar or nuclear energy to power a robot.*)
4. What type of actuators can be used in a robot? (*Gears, servo motors, hydraulic systems, and pneumatic systems are all examples of actuators that can be used in robots.*)

Effectors: How Robots Do Things, pp. 55-59

1. What are several examples of effectors? (*An example of an effector would be a gripper, tool, light, speaker, or arm.*)
2. What does "degree of freedom" refer to? (*A "degree of freedom" refers to the number of directions in which a robot part can move.*)
3. What does each degree of freedom need? (*Every degree of freedom needs a joint and an actuator to move the joint.*)

Want More

- ❒ **Mini-walker –** Have the students build their own passive dynamic walker using the activity suggested on pp. 52-54 of *Robotics*.
- ❒ **Robotic Hand –** Have the students make their own models of a robotic hand by doing the activity suggested on pp. 60-61 of *Robotics*.

Sketch Week 32

Anatomy of a Solar Cell

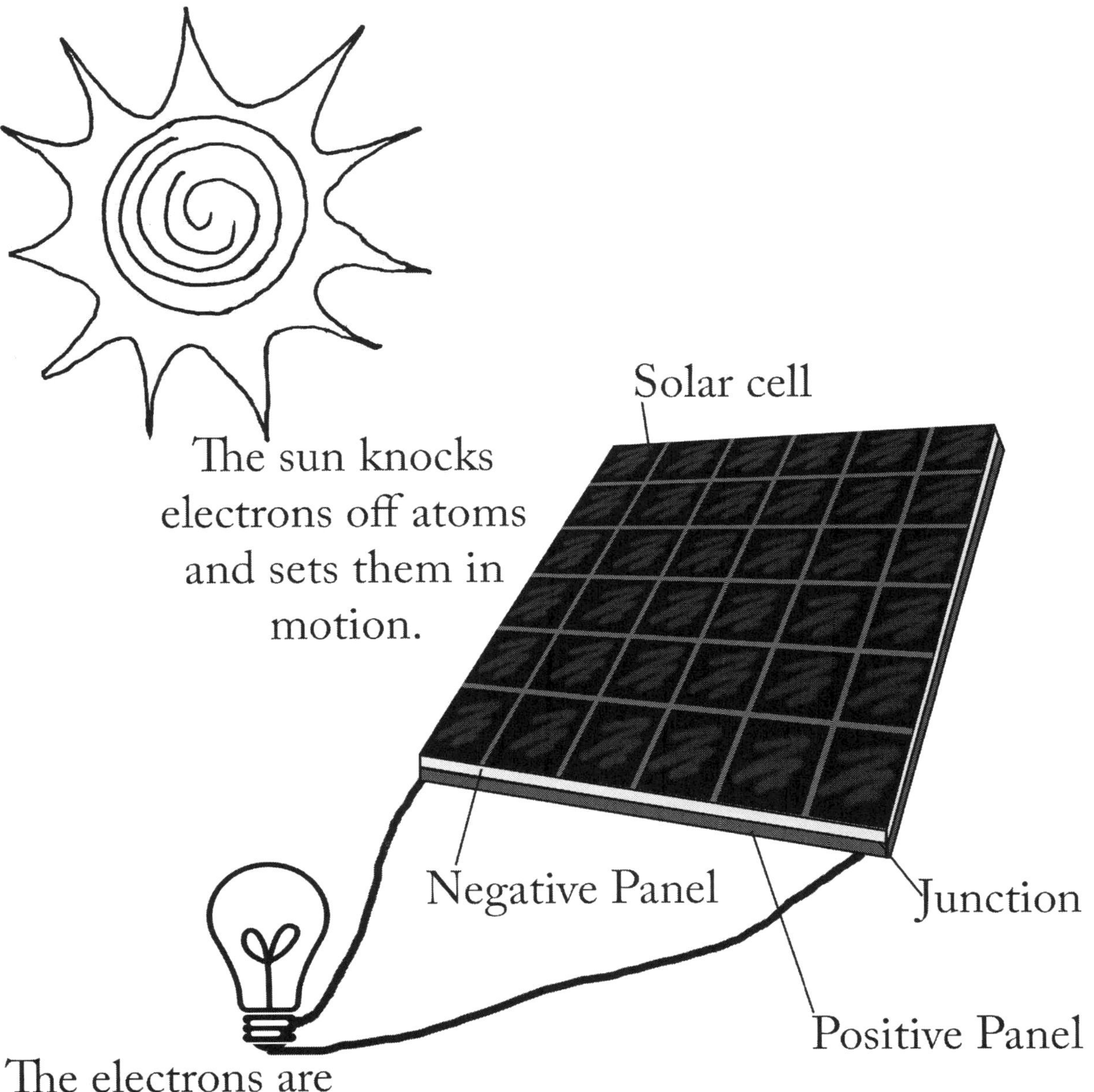

The electrons are used to power a device.

Student Assignment Sheet Week 33
Sensors and Controllers

Experiment: Pressure Sensor

Materials

- ✓ LED light bulb with two metal legs
- ✓ 3-volt Watch battery
- ✓ 2 Index cards
- ✓ Aluminum foil
- ✓ Scissors
- ✓ Marker
- ✓ Yarn
- ✓ Glue
- ✓ Toothpick
- ✓ Tissue

Procedure

1. Read the introduction to the experiment on pg. 80 of *Robotics*.
2. Follow the directions on pg. 80-81 of *Robotics* to build a pressure sensor. Then, test how the sensor works.
3. Draw conclusions and complete the experiment sheet.

Vocabulary & Memory Work

- ☐ Vocabulary: controller, photoresistor, sensor
- ☐ Memory Work—There is no memory work for this week.

Sketch: Anatomy of a Light Sensor

- Label the following – electrodes, photoconductive material, the sensor receives light input, the sensor sends information to processor.

Writing

- Reading Assignment: *Robotics* pp. 68-78 Sensors: How Robots Know What's Going On, pp. 85-93 Controller: How Robots Think
- Additional Research Readings
  - Social Robots: *Robotics* pp. 104-113

Dates

- 1947 – The transistor is invented by John Bardeen, Walter Brattain, and William Shockley. This invention makes it possible to design modern computers and robots.
- 1967 – Seymor Papert, an MIT mathematician, develops Logo, one of the earliest computer languages for children.

Schedules for Week 33

Two Days a Week

Day 1	Day 2
☐ Build the "Pressure Sensor," and then fill out the experiment sheet on SG pp. 236-237 ☐ Define controller, photoresistor, and sensor on SG pg. 203 ☐ Enter the dates onto the date sheets on SG pp. 8-13	☐ Read the Chapter 5 and 6 of *Robotics,* and then discuss what was read ☐ Color and label the "Anatomy of a Light Sensor" sketch on SG pg. 235 ☐ Prepare an outline or narrative summary; write it on SG pp. 238-239

Supplies I Need for the Week

✓ LED light bulb with two metal legs, 3-volt Watch battery,
✓ 2 Index cards, Aluminum foil, Scissors, Marker, Yarn,
✓ Glue, Toothpick, Tissue

Things I Need to Prepare

Five Days a Week

Day 1	Day 2	Day 3	Day 4	Day 5
☐ Build the "Pressure Sensor," and then fill out the experiment sheet on SG pp. 236-237 ☐ Enter the dates onto the date sheets on SG pp. 8-13	☐ Read the Chapter 5 and 6 of *Robotics,* and then discuss what was read ☐ Write an outline on SG pg. 238	☐ Define controller, photoresistor, and sensor on SG pg. 203 ☐ Color and label the "Anatomy of a Light Sensor" sketch on SG pg. 235	☐ Read one or all of the additional reading assignments ☐ Write a report on what you learned on SG pg. 239	☐ Complete one of the Want More Activities listed **OR** ☐ Study a scientist from the field of Physics

Supplies I Need for the Week

✓ LED light bulb with two metal legs, 3-volt Watch battery,
✓ 2 Index cards, Aluminum foil, Scissors, Marker, Yarn,
✓ Glue, Toothpick, Tissue

Things I Need to Prepare

Additional Information Week 33

Experiment Information

- **Introduction –** (*from the Student Guide*) Read the introduction on pressure sensors found on pg. 80 of *Robotics*.
- **Results –** The students should see the light bulb light up when they squeeze the index cards together. The amount that the light bulb shines will depend upon the pressure the students exert.
- **Explanation –** When the students squeeze the cards together, the electricity flows from the battery through the foil to the light bulb and back to the battery, allowing the circuit to be completed. The yarn prevents the cards from completely closing, allowing the students to use pressure to control the amount of electricity that flows through the circuit.

Discussion Questions

Sensors: How Robots Know What's Going On, pp. 68-78

1. How do a robot's sensors work? (*A robot's sensors take in information and convert it into an electrical signal that the robot's processor can understand.*)
2. What different types of sensors can robots have? (*Robots can have many different kinds of sensors, like ones that sense light, tilt, motion, sound, radio waves, or location.*)

Controller: How Robots Think, pp. 85-93

1. What is a micro-controller? (*A micro-controller is a small, simple computer that can be used to control a robot.*)
2. How are computers and code used in robotics? (*Computers act as brains for robots; the code contains the instructions for the robots to process information and act.*)
3. What is the most important part of computer programming? (*When writing a computer program, the most important part is to break down each and every step you want the computer to do.*)
4. What is Boolean logic? (*Boolean logic is a way to turn a decision into a yes or no question that a computer can use to process information.*)

Want More

- **Tilt Sensor –** Have the students build their own mini-ball tilt sensors using the activity suggested on pp. 72-73 of *Robotics*.
- **Binary Bead Jewelry –** Have the students create bracelets that contain the code for their names using the directions found on pp. 101-103 *Robotics*.

Sketch Week 33

Anatomy of a Light Sensor

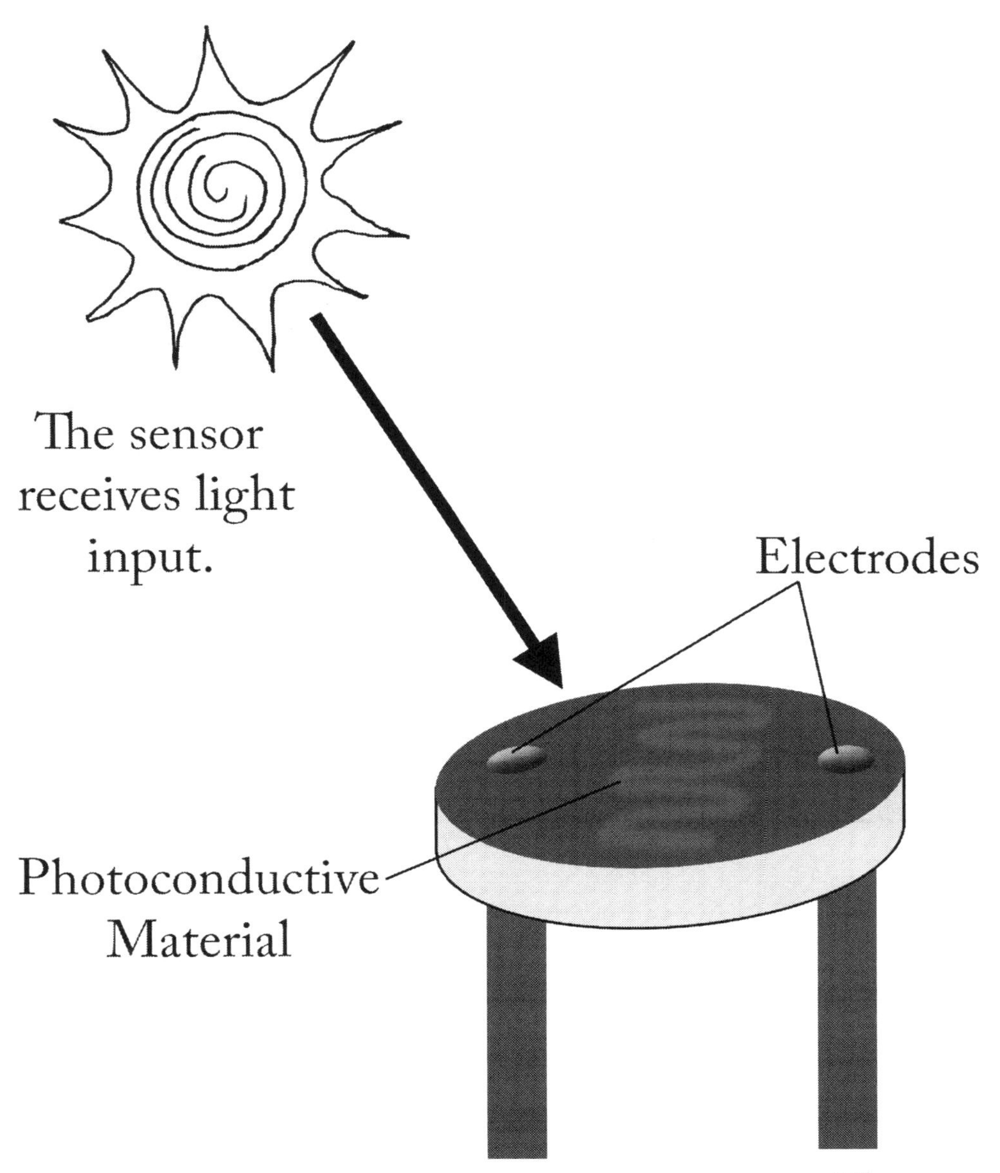

The sensor sends information to the processor.

Unit 7 Engineering and Robotics
Unit Test Answers

Vocabulary Matching

1. C
2. L
3. B
4. H
5. D
6. F
7. A
8. I
9. K
10. O
11. M
12. E
13. N
14. J
15. G

True or False

1. True
2. False (*Torsion is a force that twists and turns a material. Shear is a force that slips parts of a material in opposite directions.*)
3. True
4. True
5. True
6. False (*Natural tunnels can be dug out by animals or formed by lava.*)
7. False (*Nanobots are robots that are too small to see without a microscope.*)
8. True
9. False (*The addition of a capacitor in a BEAM robot allows the machine to store power until there is enough to make it move.*)
10. True
11. False (*A robot's sensors take in information and convert it into an electrical signal that the robot's processor can understand.*)
12. True

Short Answer

1. The main branches of engineering are chemical, civil, computer, electrical, and mechanical.
2. The suspension bridge has anchorage points at either end where cables are attached. These cables are strung across and down to hold the roadway underneath. These bridges are extremely sturdy and can span quite long distances.
3. The three stages of man-made tunnel building are excavation, support, and lining. In the excavation stage, the engineers bore, or hollow out, the passageway. In the support stage, the engineers build supports that prevent collapses and water seepage. In the lining stage, the engineers line the tunnel to provide a permanent support.
4. A biomimetic robot is a robot that copies a living thing in some way.
5. Every degree of freedom needs a joint and an actuator to move the joint.
6. Boolean logic is a way to turn a decision into a yes or no question that a computer can use to process information.
7. The Engineering Design Process:
    - Identify the Need or Problem
    - Research and Define Requirements
    - Brainstorm for Solutions
    - Design a Plan

- Build a Prototype
- Test and Evaluate the Prototype
- Communicate the Results
- Redesign if Needed

Unit 7 Engineering and Robotics
Unit Test

Vocabulary Matching

1. Engineer ____
2. Load ____
3. Beam ____
4. Bridge ____
5. Truss ____
6. Arch ____
7. Tunnel ____
8. Automaton ____
9. Robot ____
10. Actuator ____
11. Capacitor ____
12. Effector ____
13. Controller ____
14. Photoresistor ____
15. Sensor ____

A. A passageway that goes under or through a natural or man-made obstacle, such as rivers, mountains, building, roads, and more.

B. A horizontal, weight-bearing, rigid structure that carries a load.

C. A person who uses science and math to design, build, and maintain engines, machines, bridges, tunnels, and other public works structures.

D. A framework of beams of bars that can support structures, like bridges.

E. A device that lets a robot affect things in the world around it, such as grippers, tools, and laser beams.

F. A curved weight-bearing structure that is in the shape of an upside-down U.

G. A device in robotics that takes in information from the outside world.

H. A man-made structure that has been built to span rivers, canyons, roads, railways, and more.

I. A machine that can move by itself.

J. A light sensor that is able to change the resistance in an electrical current depending upon the amount of light it receives.

K. A machine that is able to sense, think, and act on its own.

L. An applied force or weight.

M. An electrical component that stores electrical charge and releases the charge when it is needed.

N. A computer or switch that can react to the information gathered by sensors.

O. A piece of equipment that makes a robot move.

True or False

1. ____________________ Tension is a force that pulls or stretches a material outward. Compression is a force that squeezes or presses a material inward.

2. ____________________ Shear is a force that twists and turns a material. Torsion is a force that slips parts of a material in opposite directions.

3. ____________________ The three major bridge designs are beam, arch, and suspension.

4. ____________________ The keystone locks the stones in the arch in place.

5. ____________________ The three tunnel types are soft-ground, hard-rock, and underwater.

6. ____________________ Natural tunnels are typically made by man.

7. ____________________ Nanobots are robots tiny robots that can be easily seen with the naked eye.

8. ____________________ Robots are used in homes, toys, art, medicine, industry, military, and space.

9. ____________________ The addition of a capacitor in a BEAM robot causes more electricity to be lost from the machine.

10. ____________________ A "degree of freedom" refers to the number of directions in which a robot part can move.

11. ____________________ A robot's sensors put out in information.

12. ____________________ Computers act as brains for a robots; the code contains the instructions for the robot to process information and act.

Short Answer

1. What are the five main branches of engineering?

2. How is a suspension bridge constructed and when is it useful?

3. What are the three stages of man-made tunnel building? Explain what happens in each.

4. What is a biomimetic robot?

5. What does each degree of freedom in a robot need?

6. What is Boolean logic?

7. What are the steps in the engineering design process?

Physics Unit 8

Nuclear Physics

Unit 8 Nuclear Physics
Overview of Study

Sequence of Study

Week 34: Radioactivity
Week 35: Nuclear Energy

Materials by Week

Week	Materials
34-35	*No experiment supplies needed.*

Note – *There are no experiments in this unit; instead, the students will be completing a scientist biography report project.*

Vocabulary for the Unit

1. **Radiation** – A stream of particles from a source of radiation.
2. **Radioactive decay** – The process by which a nucleus ejects particles by radiation until stability is reached.
3. **Radioisotope** – An unstable nucleus that has a different number of neutrons than a stable nucleus.
4. **Fission –** The process by which a heavy unstable nucleus is split into two lighter nuclei, which causes the release of several neutrons and large amounts of energy.
5. **Fusion –** The collision and combination of two lighter nuclei to form a heavier, more stable nucleus, releasing a large amount of energy.
6. **Nuclear reactor –** The part of a nuclear power plant where the nuclear fission reaction occurs.

Memory Work for the Unit

Types of Radioactive Particles

1. **Alpha Particles** – Positively charged particles that can be ejected from some radioactive nuclei. Each of the particles consist of two protons and two neutrons.
2. **Beta Particles** – Particles that are ejected from a radioactive nucleus at the speed of light. Each particle has the same mass as an electron.
3. **Gamma Rays** – Electromagnetic waves that have high penetrating power and are generally emitted from a radioactive nucleus after alpha and beta particles.

Notes

Student Assignment Sheet Week 34
Radioactivity

Scientist Biography Report Project

This week, you will complete step one and two of your Scientist Biography Report Project. You will be choosing the scientist you would like to learn more about and do your research. The instructions for this week's assignments are on the following Scientist Biography Report sheets.

Vocabulary & Memory Work

- ☐ Vocabulary: radiation, radioactive decay, radioisotope
- ☐ Memory Work—This week, begin to work on memorizing the types of radioactive particles.
  - **Types of Radioactive Particles**
    1. **Alpha Particles** – Positively charged particles that can be ejected from some radioactive nuclei. Each of the particles consist of two protons and two neutrons.
    2. **Beta Particles** – Particles that are ejected from a radioactive nucleus at the speed of light. Each particle has the same mass as an electron.
    3. **Gamma Rays** – Electromagnetic waves that have high penetrating power and are generally emitted from a radioactive nucleus after alpha and beta particles.

Sketch: Radioactive Penetrating Power

- Label the following – alpha particle, beta particle, gamma rays, thick sheet of paper, sheet of aluminum, thick sheet of lead
- Draw the arrows show the penetrating abilities of the rays.

Writing

- Reading Assignment: *DK Encyclopedia of Science* pp. 26-27 Radioactivity
- Additional Research Readings
  - Radiation: *KSE* pp. 244-245
  - Radioactivity: *UIDS* pp. 86-87
  - Uses of Radioactivity: *UIDS* pg. 91

Dates

- 1896 – Antoine Becquerel is working with a natural fluorescent material and x-rays, which leads to his discovery of radioactivity.
- 1898 – Marie and Pierre Curie coin the term "radioactive" when they discovered radium and polonium.
- 1928 – Hans Geiger and one of his students develop a machine that can detect and measure the intensity of radiation.

Schedules for Week 34

Two Days a Week

Day 1	Day 2
☐ Choose your scientist and do your research for the Scientist Biography Report,see SG pp. 246-247 for details ☐ Define radiation, radioisotope, and radioactive decay on SG pg. 242 ☐ Enter the dates onto the date sheets on SG pp. 8-13	☐ Read pp. 26-27 from *DK EOS,* then discuss what was read ☐ Color and label the "Radioactive Penetrating Power" sketch on SG pg. 245 ☐ Prepare an outline or narrative summary, write it on SG pp. 248-249
Supplies I Need for the Week	
Things I Need to Prepare	

Five Days a Week

Day 1	Day 2	Day 3	Day 4	Day 5
☐ Choose your scientist and begin your research for the Scientist Biography Report,see SG pp. 246-247 for details	☐ Read pp. 26-27 from *DK EOS,* then discuss what was read ☐ Write an outline on SG pp. 248-249	☐ Define radiation, radioisotope, and radioactive decay on SG pg. 242 ☐ Color and label the "Radioactive Penetrating Power" sketch on SG pg. 245	☐ Enter the dates onto the date sheets on SG pp. 8-13 ☐ Continue working on your Scientist Biography Report	☐ Continue working on your Scientist Biography Report
Supplies I Need for the Week				
Things I Need to Prepare				

Additional Information Week 34

Scientist Biography Report Step 1: Choose a Scientist

☞ **From the Student Guide –** During the next two weeks, you are going to be researching and learning more about a scientist that has contributed to the field of physics. This week, you need to begin your scientist biography project by choosing which scientist you will research. You can choose one of the scientists mentioned in the "Dates" sections or you can choose one that has interested you.

☞ **Step 1 Notes –** The students will need some help with step one, especially if they are not familiar with some of the well-known scientists in the field of physics. They can choose one of the more famous physicists, like Albert Einstein, or one of the lesser known physicists, such as James Joule. Either way, they need to choose a person that has enough information written about him or her for the students to compile a two to three page paper. Here are a few options:

1. Albert Einstein
2. Max Born
3. Isaac Newton
4. Marie Curie
5. Neils Bohr
6. Richard Feynman
7. Steven Hawking
8. Michael Faraday
9. Galileo Galilei
10. James Maxwell
11. Max Planck
12. Nikola Tesla
13. James Joule
14. Johannes Kepler

Scientist Biography Report Step 2: Research the Scientist

☞ **From the Student Guide –** Once you have chosen the scientist you would like to study, you can begin your research. Begin by looking for a biography on your chosen scientist at the library. Then, look for articles on the physicist in magazines, newspapers, encyclopedias, or on the Internet. You will need to know the following about your scientist to write your report:

☑ Biographical information on the scientist (*i.e., where they were born, their parents, siblings, and how they grew up*);

☑ The scientist's education (*i.e., where they went to school, what kind of student they were, what they studied, and so on*);

☑ Their scientific contributions (*i.e., research that they participated in, any significant discoveries they made, and the state of the world at the time of their contributions*).

As you read over the material you have gathered, be sure to write down any facts you glean in your own words. You can do this on the sheet below or on separate index cards.

☞ **Step 2 Notes –** Students are strongly recommended to use the index card system to record their research findings. You can read more about this method by clicking below:

💻 http://elementalblogging.com/the-index-card-system/

Discussion Questions

1. What is meant by a radioactive atom? (*A radioactive atom has an unstable nucleus that can easily split up.*)
2. How is the number of subatomic particles related to an atom's potential radioactivity? (*The*

larger number of subatomic particles an atom has, the more likely it is to be radioactive.)

3. Why are radioactive materials often stored in water? (*Radioactive materials are often stored in water because the water absorbs the radiation and acts as a shield.*)
4. What are some helpful uses of radiation? (*Radiation is used in pacemaker batteries, cancer treatments, and smoke alarms.*)
5. What is radioactive fallout? (*Radioactive fallout is when radioactive material is thrown out into the air from an explosion or accident and the material returns to the ground as fallout.*)
6. What is a half-life? (*A half-life is the time it takes for half the atoms in a radioactive substance to take the first decay step. It is also known as the rate of radioactive decay.*)

Want More

- **Radioactive Decay** – Have the students do the "Alpha Decay" simulation from PhET. Here is the link to the on-line demonstration:
  - http://phet.colorado.edu/en/simulation/alpha-decay
- **Carbon Dating** – Have the students research the role of radioactivity in carbon dating. After they have gathered enough information, have the students write a short report sharing what they have learned.

Sketch Week 34

Radioactive Penetrating Power

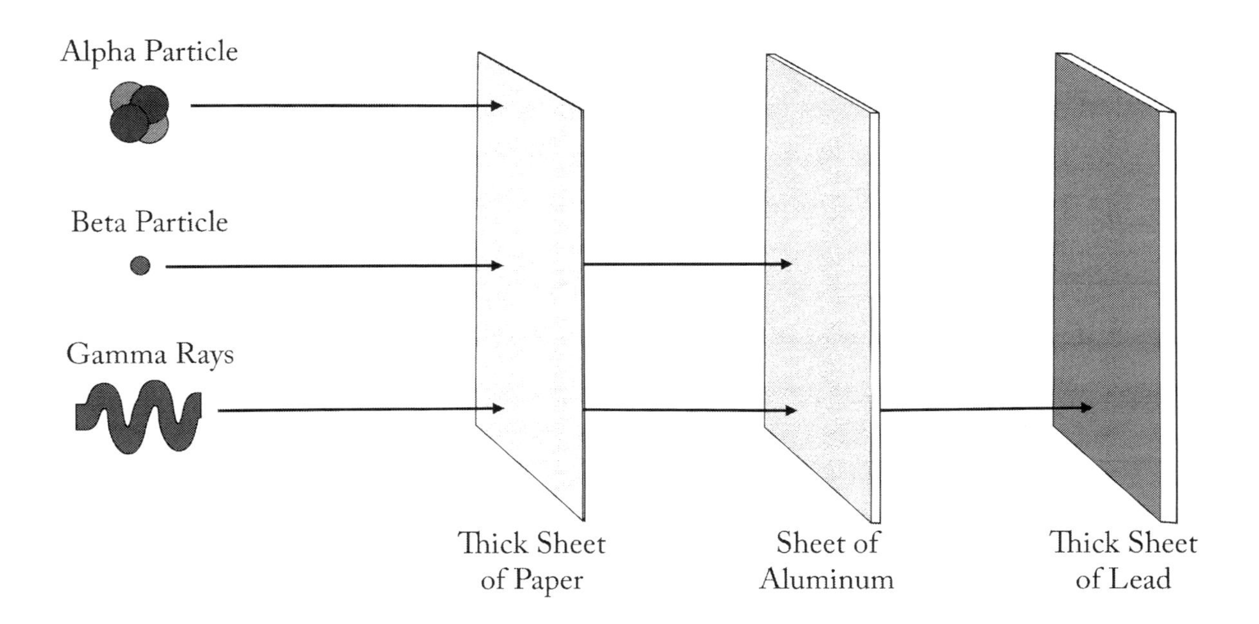

Student Assignment Sheet Week 35
Nuclear Energy

Scientist Biography Report Project

This week, you will complete step three through five of the Scientist Biography Report Project. You will be finishing up your research and organizing the information you have gathered into an outline. The instructions for this week's assignments are on the following Scientist Biography Report sheets.

Vocabulary & Memory Work

- ☐ Vocabulary: fission, fusion, nuclear reactor
- ☐ Memory Work—This week, continue to work on memorizing the types of radioactive particles.

Sketch: Nuclear Fission

- Label the following – high speed neutron, unstable nucleus, energy, neutrons, nuclei

Writing

- Reading Assignment: *DK Encyclopedia of Science* pp. 136-137 Nuclear Energy
- Additional Research Readings
  - Power Plants (section on Nuclear Power): *KSE* pp. 348-349
  - Atomic and Nuclear Energy: *UIDS* pp. 84-85
  - Nuclear fission and fusion: *UIDS* pp. 92-93

Dates

- 1919 – Ernst Rutherford changes the nucleus of a nitrogen atom into an oxygen nucleus.
- 1956 – The first nuclear power plant, located in England, starts generating power.
- 1986 – The nuclear reactor at Chernobyl melts down, throwing radioactive material into the air.

Schedules for Week 35

Two Days a Week

Day 1	Day 2
☐ Write the outline and rough draft of the Scientist Biography Report, see SG pp. 252-254 for details ☐ Define fission, fusion, and nuclear reactor on SG pg. 242 ☐ Enter the dates onto the date sheets on SG pp. 8-13	☐ Revise and write the final draft of the Scientist Biography Report, see SG pg. 255 for details ☐ Read pp. 136-137 from *DK EOS,* then discuss what was read ☐ Color and label the "Nuclear Fission" sketch on SG pg. 251 ☐ Prepare an outline or narrative summary, write it on SG pp. 256-257
Supplies I Need for the Week	
Things I Need to Prepare	

Five Days a Week

Day 1	Day 2	Day 3	Day 4	Day 5
☐ Write the outline for the Scientist Biography Report, see SG pp. 252-254 for details ☐ Enter the dates onto the date sheets on SG pp. 8-13	☐ Write the rough draft for the Scientist Biography Report	☐ Revise your paper and write the final draft for the Scientist Biography Report, see SG pg. 255 for details	☐ Read pp. 136-137 from *DK EOS,* then discuss what was read ☐ Write an outline on SG pp. 256-257	☐ Define fission, fusion, and nuclear reactor on SG pg. 242 ☐ Color and label the "Nuclear Fission" sketch on SG pg. 251
Supplies I Need for the Week				
Things I Need to Prepare				

Additional Information Week 35

Scientist Biography Report Step 3: Create an Outline

☞ **From the Student Guide –** Now that your research is completed, you are ready to begin the process of writing a report on your chosen scientist. This week, you are going to organize the notes you took during step two into a formal outline which you will use next week to write the rough draft of your report. Use the outline template provided on the student sheets as a guide. You should include information on why you chose the particular scientist in your introduction section. For the conclusion section of the outline, you need to include why you believe someone else should learn about your chosen scientist and your impression of the scientist (i.e., *Did you like the scientist? Do you feel that they made a significant impact on the field of physics?*).

☞ **Step 3 Notes –** The outline the students create can look like the one below.

Scientist Biography Outline

I. Introduction and Biological Information on the Scientist
(4-6 points)
II. The Scientist's Education
(4-6 points)
III. The Scientist's Contributions
(1-3 sub categories each with 4-5 points)
IV. Conclusion
(4-5 points)

Scientist Biography Report Step 4: Write a Rough Draft

☞ **From the Student Guide –** Last week, you created a formal outline for your scientist biography report; now, it is time to take that outline and turn it into a rough draft. Simply take the points on your outline, combine, and add in sentence openers to create a cohesive paragraph. Here's what your rough draft should look like:

- ☑ Paragraph 1 (*from outline point I*): introduce the scientist;
- ☑ Paragraph 2 (*from outline point II*): tell about the scientist's education;
- ☑ Paragraph 3-5 (*from outline point III*): share the scientist's contributions (*one paragraph for each contribution*);
- ☑ Paragraph 6 (*from outline point IV*): share your thoughts on the scientist and why someone should learn about him or her.

You can choose to hand write or type up your rough draft on a separate sheet of paper. However, keep in mind that you will need a typed version for step five.

Scientist Biography Report Step 5: Revise to Create a Final Draft

☞ **From the Student Guide –** Now that you have a typed, double-spaced rough draft, look over it one more time to make any changes you would like. Then, have your teacher or one of your peers look over the paper for you to correct any errors and bring clarity to any of the unclear sections. Once this is complete, make the necessary changes to your paper to create your final draft. Print out your final paper and include it on the next page.

☞ **Step 5 Notes –** A grading rubric for this assignment is included in the Appendix on pp. 267-

268. If you would like to change things up for your student, have them create a poster or mini-book for the final draft of their Scientist Biography Report.

Discussion Questions

1. What is the basic difference between nuclear fission and fusion? (*In fission, the atoms split, but in fusion the atoms come together.*)
2. What elements are typically used in nuclear fuel rods? (*Nuclear fuel rods are typically made of uranium or uranium dioxide.*)
3. Why are boron rods found in a nuclear reactor? (*Boron rods can be found in a nuclear reactor because the rods are able to absorb neutrons and help control the rate of the nuclear reaction.*)
4. What is the difference between an atomic bomb and a hydrogen bomb? (*An atomic bomb uses uncontrolled nuclear fission, while a hydrogen bomb uses nuclear fusion.*)
5. Why is fusion not a practical method of producing power on Earth yet? (*Fusion is not a practical method of producing power yet because the plasma, or gases, that are fused need to be heated to an extremely high temperature.*)

Want More

- **Nuclear Fission** – Have the students do the "Nuclear Fission" simulation from PhET. Here is the link to the on-line demonstration:
  - http://phet.colorado.edu/en/simulation/nuclear-fission

Sketch Week 35

Nuclear Fission

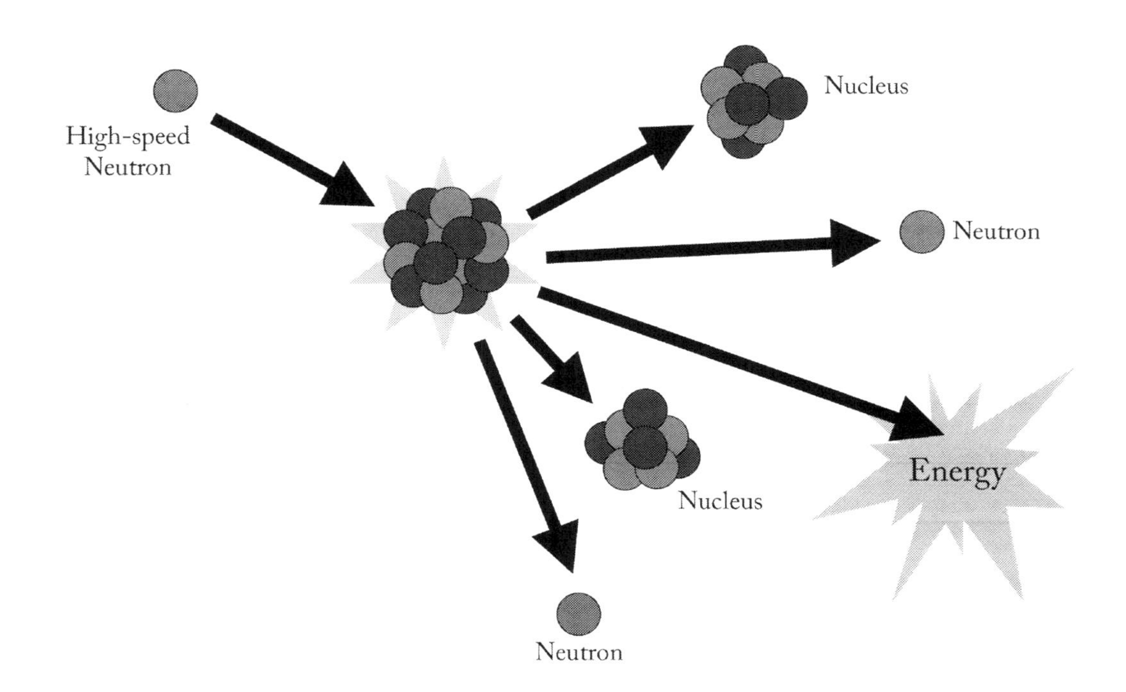

Unit 8 Nuclear Physics
Unit Test Answers

Vocabulary Matching

1. B
2. F
3. D
4. A
5. C
6. E

True or False

1. True
2. True
3. True
4. False (*An atomic bomb uses uncontrolled nuclear fission, while a hydrogen bomb uses nuclear fusion.*)

Short Answer

1. In fission, the atoms split, but in fusion the atoms come together.
2. A half-life is the time it takes for half the atoms in a radioactive substance to take the first decay step. It is also known as the rate of radioactive decay.
3. Types of Radioactive Particles
   - **Alpha Particles** – Positively charged particles that can be ejected from some radioactive nuclei. Each of the particles consist of two protons and two neutrons.
   - **Beta Particles** – Particles that are ejected from a radioactive nucleus at the speed of light. Each particle has the same mass as an electron.
   - **Gamma Rays** – Electromagnetic waves that have high penetrating power and are generally emitted from a radioactive nucleus after alpha and beta particles.

Unit 8 Nuclear Physics
Unit Test

Vocabulary Matching

1. Radiation ___
2. Radioactive Decay ___
3. Radioisotopes ___
4. Fission ___
5. Fusion ___
6. Nuclear reactor ___

A. The process by which a heavy unstable nucleus is split to two lighter nuclei, which causes the release of several neutrons and large amounts of energy.

B. A stream of particles from a source of radiation.

C. The collision and combination of two lighter nuclei to form a heavier, more stable nucleus, releasing a large amount of energy.

D. An unstable nucleus that has a different number of neutrons than a stable nucleus.

E. The part of a nuclear power plant where the nuclear fission reaction occurs.

F. The process by which a nucleus ejects particles through radiation until stability is reached.

True or False

1. ______________________ A radioactive atom has an unstable nucleus that can easily split up.

2. ______________________ Radiation is used in pacemaker batteries, cancer treatments, and smoke alarms.

3. ______________________ Nuclear fuel rods are typically made of uranium or uranium dioxide.

4. ______________________ A hydrogen bomb uses uncontrolled nuclear fission, while an atomic bomb uses nuclear fusion.

Short Answer

1. What is the basic difference between nuclear fission and fusion?

2. What is a half-life?

3. What are the three types of radioactive particles?

Physics: Wrap-up

Year-end Test

Physics for the Logic Stage
Year-end Test Information and Answers

Year-end Test Information

The year-end test is based on the vocabulary, short answer questions from the tests, and memory work throughout the year. You can choose to make it open notes or not. The purpose of this test is to help your students gain familiarity with the concept of a final exam, so that it won't be quite so overwhelming when they reach the high school years.

Year-end Test Answers

Vocabulary Matching

1. C
2. H
3. A
4. E
5. G
6. D
7. I
8. K
9. L
10. N
11. P
12. F
13. B
14. M
15. R
16. J
17. O
18. Q
19. V
20. X
21. W
22. T
23. U
24. Y
25. AA
26. Z
27. AF
28. AB
29. AE
30. S
31. AG
32. AD
33. AI
34. AC
35. AJ
36. AH

Multiple Choice

1. C
2. A
3. D
4. B
5. C, F, B, A, D, G, E
6. A
7. E
8. 3, 2, 4, 1
9. D
10. B
11. B, D, C, A
12. A, C, E
13. A
14. C
15. B
16. C, A, B
17. 2, 3, 1
18. B, C, A

Short Answer

1. Newton's Three Laws of Motion
   - An object will not move unless a force, like a push or pull, moves it. Once it is moving, an object will not stop moving in a straight line unless it's forced to change.
   - The greater the force on an object, the greater the change in its motion. The greater the mass of an object, the greater the force needed to change its motion.
   - For every reaction, there is an equal but opposite reaction.
2. Kinetic energy is the energy of motion, but potential energy is the energy that an object has because of its position.

3. Heat can travel by conduction, convection, and radiation. Conduction is how heat travels through a solids. Convection is how heat travels through liquids and gases. Radiation is how heat travels through empty spaces.
4. Sound travels more quickly through a solid because the molecules are closer together and more firmly in place. Therefore, they bounce back into place quicker than the molecules in a gas.
5. Quantum theory suggests that light has a combination of the properties of a wave and of a particle.
6. Attraction is when two objects with opposite electrical charges move towards one another. Repulsion is when two objects with the same electrical charge move away from one another.
7. The Earth's magnetic field is caused by the iron core at the center of the Earth. The Earth's magnetic field allows us to use a compass to navigate. It also cause an aurora to be produced, known as the northern lights.
8. Every degree of freedom needs a joint and an actuator to move the joint.
9. A half-life is the time it takes for half the atoms in a radioactive substance to take the first decay step. It is also known as the rate of radioactive decay.

Physics for the Logic Stage
Year-end Test

Vocabulary Matching

1. Force ___
2. Inertia ___
3. Mass ___
4. Acceleration ___
5. Velocity ___
6. Energy ___
7. Work ___
8. Pressure ___
9. Input force ___
10. Output force ___
11. Energy conversion ___
12. Entropy ___
13. Heat ___
14. Power ___
15. Mechanical wave ___
16. Sound ___
17. Amplitude ___

A. The amount of matter in an object.

B. A form of energy that flows from a place of high temperature to a place of lower temperature.

C. A push or pull that acts on an object.

D. The ability to do work.

E. A change in an object's speed, direction, or both.

F. The degree of disorder in a given system.

G. The speed of an object in a particular direction.

H. The tendency of an object to resist a change in its motion.

I. The transfer of energy that occurs when a force moves or changes an object.

J. A mechanical wave, or vibration, that travels through a medium, such as air, and can be heard when it reaches a person's or animal's ear.

K. The amount of force pushing on a giving area.

L. The force you put into a machine.

M. The rate at which work is done or energy is used.

N. The force that a machine exerts on an object.

O. The size of a vibration or the height of a wave.

P. The process of changing one form of energy into another.

Q. The number of waves that pass a given point in a second.

R. A wave that travels through a medium, such as air, water, or solids.

S. An applied force or weight.

T. A piece of transparent material with a curved edge that causes light to bend in a particular way.

18. Frequency ___
19. Wavelength ___
20. Electromagnetic waves ___
21. Photons ___
22. Lens ___
23. Conductor ___
24. Insulator ___
25. Resistance ___
26. Magnetic field ___
27. Magnetic pole ___
28. Electromagnetic force ___
29. Engineer ___
30. Load ___
31. Actuator ___
32. Capacitor ___
33. Sensor ___
34. Radioactive decay ___
35. Fission ___
36. Fusion ___

U. A material through which electrical charge can easily flow.

V. The distance between the crest of one wave to the crest of another.

W. Packets of electromagnetic energy.

X. These waves carry energy from one place to another and are produced when an electric charge vibrates or accelerates.

Y. A material through which electrical charge cannot easily flow.

Z. The area around a magnet in which the magnetic force can be felt.

AA. The ability of a material to resist the flow of electrical current.

AB. The force that is produced when an electrical current flows through a wire, which forms a magnetic field.

AC. The process by which a nucleus ejects particles through radiation until stability is reached.

AD. An electrical component that stores electrical charge and releases the charge when it is needed.

AE. A person who uses science and math to design, build, and maintain engines, machines, bridges, tunnels, and other public works structures.

AF. One of the two ends of a magnet where the force of attraction or repulsion is the strongest.

AG. A piece of equipment that makes a robot move.

AH. The collision and combination of two lighter nuclei to form a heavier, more stable nucleus, releasing a large amount of energy.

AI. A device in robotics that takes in information from the outside world.

AJ. The process by which a heavy unstable nucleus is split into two lighter nuclei, which causes the release of several neutrons and large amounts of energy.

Multiple Choice

1. What two things affect the force of gravity between two objects?

   A. Only the distance between the objects affects the force of gravity.

   B. Only the mass of the objects both affects the force of gravity.

   C. Both the distance between the objects and the mass of the objects affect the force of gravity.

2. When is an object in balance?

   A. An object is balanced when the forces acting on it add to each other, producing a zero resultant force.

   B. An object is balanced when the forces acting on it add to each other, producing a negative resultant force.

   C. An object is balanced when the forces acting on it add to each other, producing a positive resultant force.

3. Which of the following is true about friction?

   A. A streamlined design reduces the friction felt by an object.

   B. The rougher the surface, the more friction is produced.

   C. Friction is what opposes the motion of objects that touch as they move past each other.

   D. All of the above.

4. Which of the following statements are NOT true about the relationship between pressure, force, and area.

   A. If you spread force over a large area, you will reduce the amount of pressure exerted.

   B. If you spread a force over a large area, you will increase the amount of pressure exerted.

   C. If you concentrate the force over a small area, you will increase the amount of pressure exerted.

4. Match the following simple machines with their description.

Lever ___

Wheel and axle ___

Gears ___

Inclined plane ___

Wedge ___

Screw ___

Pulley ___

A. A slanted surface that helps move objects up an incline.

B. Toothed wheels that interlock in pairs; each one helps to drive the next.

C. A rigid bar that is free to move around at a fixed point.

D. A v-shaped object whose sides are two inclined planes.

E. A rope that fits in the groove of a wheel.

F. Two disks or cylinders, each with a different radius.

G. An inclined plane wrapped around a cylinder.

5. How are heat and temperature related?

   A. Heat is the energy that an object has because the molecules are moving, while temperature is the measure of how fast the object's molecules are moving.

   B. Heat is the measure of how fast the object's molecules are moving, while temperature is the energy that an object has because the molecules are moving.

6. Which of the following is NOT a law of thermodynamics?

   A. When two systems are in equilibrium with a third system, they are said to be in thermal equilibrium with each other.

   B. Energy cannot be created or destroyed.

   C. Disorder (entropy) in the universe is always increasing.

   D. There is a theoretical point at which all molecular movement stops, which is known as absolute zero.

   E. None of the above.

7. Put the steps of how we hear sound in order.

   ___ In the inner ear, fluid carries the vibrations through a narrow tube that stimulates the hairs lining the tube.

   ___ Our outer ears collect sound waves, which cause the eardrum to vibrate.

   ___ This causes nerves to send electrical impulses to the brain, which we recognize as sound.

   ___ The vibrations are then carried into the inner ear by a series of tiny bones.

8. What is true about sound waves?

   A. When sounds waves hit a hard surface, the wave are reflected and bounce back.

   B. When sound waves hit a soft surface, some or all of the waves are absorbed.

   C. The loudness of a sound is determined by the amplitude of the sound wave.

   D. All of the above.

9. What does the law of reflection state?

   A. The law of reflection states that the angle of reflection is greater than the angle of incidence.

   B. The law of reflection states that the angle of reflection is equal to the angle of incidence.

   C. The law of reflection states that the angle of reflection is less than the angle of incidence.

10. Match the following terms with their definitions.

Reflection ___

Refraction ___

Convex lens ___

Concave lens ___

A. A lens that curves inward.

B. The bouncing back of light from a surface.

C. A lens that curves outward.

D. The change in direction of a light beam as it passes from one medium to another of different density.

11. Circle the three key components of a cell.

   A. A positive electrode

   B. A copper wire

   C. A negative electrode

   D. An aluminum casing

   E. An electrolyte chemical

12. Which statement is true about diodes?

   A. The anode is positively charged, the cathode is negatively charged.

   B. The cathode is positively charged, the anode is negatively charged.

13. What is a benefit of an electromagnet?

A. Electromagnets can be turned off and on.

B. Electromagnets can alter the attractive strength.

C. Both of the above.

14. The north pole of one magnet is always attracted to the north pole of another magnet.

A. True

B. False

15. Match the three main types of bridges with their description.

Beam ___

Arch ___

Suspension ___

A. This bridge has anchorage points at either end where cables are attached.

B. This bridge has a long rigid horizontal beam and two vertical abutments that support the beam at either end.

C. This bridge has an arch or a series of arches that transfers the load to the abutments.

16. Put the three stages of tunnel building in order.

___ In the support stage, the engineers build supports that prevent collapses and water seepage.

___ In the lining stage, the engineers line the tunnel to provide a permanent support for the tunnel.

___ In the excavation stage, the engineers bore, or hollow out, the passageway.

17. Match the type of radioactive particles with their description.

Alpha Particles ___

Beta Particles ___

Gamma Rays ___

A. Particles that are ejected from a radioactive nucleus at the speed of light.

B. Electromagnetic waves that have high penetrating power.

C. Positively charged particles that can be ejected from some radioactive nuclei.

Short Answer

1. What are Newton's Three Laws of Motion?

2. What is the difference between kinetic and potential energy?

3. What are the three ways heat can travel? Give a brief description of each.

4. Why does sound travel more quickly through a solid than a gas?

5. What does quantum theory say about light?

6. What are attraction and repulsion?

7. What causes the Earth's magnetic field? How does the magnetic field affect life on the surface of the Earth?

8. What does each degree of freedom in a robot need?

9. What is a half-life?

Appendix

Ancients 5000 BC-400 AD

- 6th century BC – Pythagoras directly links the amplitude of the vibration of a plucked string to the perceived loudness of the instrument.
- 585 BC – Greek philosopher, Thales, inadvertently discovers static electricity when he rubs a piece of amber with fur and observes how the amber now attracts small objects, like feathers.
- c.550 BC – Ancient philosopher, Thales Miletus, believes that there is a conservation of some sort of hidden substance of which everything is made. (Note—Today we call this hidden substance mass energy.)
- 520 BC – Greek architect, Eupalinos, becomes the first engineer.
- c330 BC – Aristotle proposes that a force is needed to maintain motion.
- c. 3rd century BC – Archimedes is said to have invented a screw pump to help get water from a reser-voir source to the fields for irrigation.
- 200 BC – The Chinese create mechanical musicians to play music for the emperor.
- 62 BC – The Ponte Fabrico Bridge, a stone double-arch bridge, is built by the Roman Empire.
- 27 BC-393 AD – The Roman Empire constructs many arch bridges, some of which still stand today.
- c. 100 – The Romans began to burn coal as a source of energy.

Medieval-Early Renaissance 400AD-1600AD

- c. 650 – The Persians began to use windmills as a source of energy.
- 1100 – Chinese sailors are the first on record for using a magnetic compass for navigating on a cloudy day.
- 1544-1603 – William Gilbert lives. He is known as the father of electricity and magnetism.
- 1555 – Gianello Torriano, an Italian clock maker, makes a model of a wind-up lady that can walk in a circle and strum a lute.

Late Renaissance-Early Modern 1600 AD-1850 AD

- 1600 – William Gilbert, an English doctor and physicist, publishes a book which for the first time explains exactly how a compass works.
- 1630's – Galileo does a series of experiments with a marble and a series of differently-shaped tracks, which leads to the discovery of a retarding force called friction.
- 1642-1727 – Isaac Newton, the English scientist who explained how force, mass, and acceleration are related, lives. The unit of force, the newton (N), is named after him.
- 1643 – Evangelista Torricelli invents the mercury barometer.

Late Renaissance-Early Modern 1600 AD-1850 AD

- 1665 – The plague breaks out in London, which forces Isaac Newton to leave Trinity College in Cambridge. He goes home and spends the next two years working on his book, Principia, in which he shares his three laws of motion.
- Late 1600's – Antoni Van Leeuwenhoek designs the first microscope.
- 1672 – Isaac Newton suggests that light is composed of tiny particles resembling balls.
- 1678 – Christian Huygens suggests that light is a wave motion, similar to sound or water waves.
- 1708 – William Derham successfully establishes the speed of sound.
- 1712 – The first steam engine is built by Thomas Newcomen.
- 1714 – German physicist, Gabriel Fahrenheit, proposes a temperature scale where the lowest point was where he could cool brine and highest point was average internal temperature of the human body. It is eventually adopted and named after him.
- 1742 – Swedish physicist, Anders Celsius, develops a temperature scale where 0 represents the freezing point of water and 100 represents the boiling point of water. It is eventually adopted and named after him.
- 1753 – Benjamin Franklin announces his lightning conductor invention, which he created as a result of his famous kite-flying experiments.
- 1765 – James Watt improves upon the original Newcomen steam engine.
- 1784-1789 – Charles Coulomb, a French physicist, writes and proves the law of electrostatics.
- 1789 – Will Herschel designs a telescope with a four foot diameter.
- 1800 – Italian scientist, Volta, invents the first battery able to hold an electrical charge.
- 1818-1889 – James Joule lives. He is credited with discovering that work produces heat, which is a form of energy.
- 1820 – Hans Christiaan Oersted is performing an experiment with electrical current when he noticed that the needle of a nearby compass moved. This realization led to the discovery of electromagnetism.
- 1821 – Michael Faraday discovers that electricity can produce rotary motion.
- 1822 – Charles Babbage designs a mechanical calculator that uses punch cards.
- 1827 – George Ohm publishes his work on resistance, including an equation that eventually becomes known as Ohm's Law.
- 1828 – Andre Ampere is elected as a member of the Royal Swedish Academy of Science in recognition of his contributions to the creation of the field of modern electrical science.
- 1840's – James Joule does a number of experiments that lead to the development of the law of conservation of energy.
- 1848 – One of the earliest wire suspension bridges is designed by Charles Ellet and built to span the Niagara River near Niagara Falls.
- 1848 – William Thomson, Lord Kelvin, develops an absolute temperature scale, known as the Kelvin scale.

Modern 1850 AD-Present

- 1850 – German scientist, Rudolf Clausius, suggests that the law of conservation of energy should be called the first law of thermodynamics.
- 1852 – William Thomson (Lord Kelvin) comes up with the idea of a "heat pump," a device that moves heat from a cold place to a hot one.
- 1860 – The first internal combustion engine is built by Etienne Lenoir. He uses coal gas and air for fuel.
- 1865 – Rudolf Clausius coins the term "entropy," which refers to unusable energy.
- 1869-1883 – The Brooklyn Bridge, a steel-cable suspension bridge, is built.
- 1877 – The first sound recording is made by Thomas Edison on a phonograph through the mechanical vibrations of a needle.
- 1884 – The first steam engine to generate electricity is invented by Charles Parsons.
- 1887 – Henrich Hertz proves that electricity can be transmitted in electromagnetic waves. He was also the first to produce and detect radio waves.
- 1890's – Ernst Mach describes how shock waves form and, along with his son Ludwig, develops a way to take pictures of the shadow of an invisible shock wave.
- 1891 – Hydroelectric power is first demonstrated in Germany.
- 1892 – James Dewar invents the vacuum flask, which is designed to prevent the transfer of heat.
- 1896 – Antoine Becquerel is working with a natural fluorescent material and x-rays, which leads to his discovery of radioactivity.
- 1898 – Marie and Pierre Curie coin the term "radioactive" when they discovered radium and polonium.
- 1899 – Alfred Nobel invents dynamite, which changes tunnel and bridge building forever.
- 1900 – Max Planck suggests that light is a combination of a particle and a wave, forming the basis of quantum theory.
- 1902 – Hendrick Lorentz is awarded the Nobel Peace Prize for his work with electromagnetic waves and the propagation of light.
- 1905 – Albert Einstein publishes his theory of relativity, which is the basis for many of the ideas we have about our universe.
- 1905 – Einstein expands upon Planck's work and suggests that light is composed of tiny particles, called photons, which have energy and behave like waves.
- 1906 – American naval architect Lewis Nixon invents the first sonar-like listening device, which he used to detect icebergs.
- 1919 – Ernst Rutherford changes the nucleus of a nitrogen atom into an oxygen nucleus.
- 1920 – Ralph Fowler develops the zeroth law of thermodynamics.
- 1926 – The first rocket propelled by liquid fuel is launched by Robert Goddard.
- 1928 – Hans Geiger and one of his students develop a machine that can detect and measure the intensity of radiation.

Modern 1850 AD-Present

- 1947 – The transistor is invented by John Bardeen, Walter Brattain, and William Shockley. This invention makes it possible to design modern computers and robots.
- 1949 – The world's first geodesic dome is built by Richard Fuller.
- 1952 – Magnets are made out of ceramics for the first time.
- 1955 – Christopher Cockerell invents the hovercraft, which uses a cushion of air that allows a vehicle to move without friction.
- 1956 – The first nuclear power plant, located in England, starts generating power.
- 1959 – The Massachusetts Institute of Technology (MIT) opens the first lab to study artificial intelligence.
- 1960 – Turkmenistan builds the first solar thermal power plant.
- 1967 – Seymor Papert, an MIT mathematician, develops Logo, one of the earliest computer languages for children.
- 1979 – Pakistani scientist, Abdus Salam, wins the Nobel Prize in Physics for his work with forces.
- 1983 – Neodymium magnets are first invented.
- 1986 – The nuclear reactor at Chernobyl melts down, throwing radioactive material into the air.
- 1987 – The Nobel Prize is awarded to Muller and Bednorz for their work with finding superconductors that function above absolute zero.
- 1989 – Mark Tilden coins the term "BEAM robot" to refer to a simple solar-powered, life-like robot.
- 1992 – The Keck telescope is built, which is thirty-three feet in diameter.
- 2000 – The Honda car company develops ASIMO, which revolutionized the way robots get around.

The Scientific Method Explained

The scientific method is a method for asking and answering scientific questions. This is done through observation and experimentation.

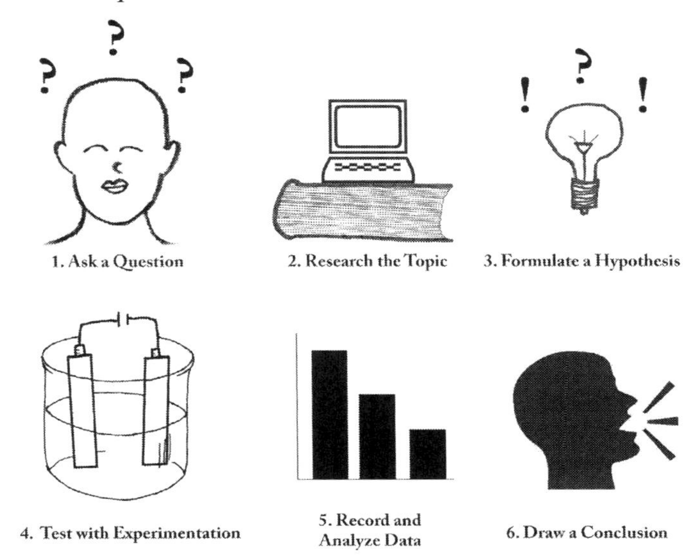

The following steps are key to the scientific method:

1. **Ask A Question** – The scientific method begins with asking a question about something you observe. Your questions must be about something you can measure. Good questions begin with how, what, when, who, which, why, or where.

2. **Do Some Research** – You need to read about the topic of your question so that you can have background knowledge of the topic. This will keep you from repeating mistakes made in the past.

3. **Formulate a Hypothesis** – A hypothesis is an educated guess about the answer to your question. Your hypothesis must be easy to measure and answer the original question you asked.

4. **Test with Experimentation** – Your experiment tests whether your hypothesis is true or false. It is important for your test to be fair. This means that you may need to run multiple tests. If you do, be sure to only change one factor at a time so that you can determine which factor is causing the difference.

5. **Record and Analyze Observations or Results** – Once your experiment is complete, you will collect and measure all your data to see if your hypothesis is true or false. Scientists often find that their hypothesis was false. If this is the case, they will formulate a new hypothesis and begin the process again until they are able to answer their question.

6. **Draw a Conclusion** – Once you have analyzed your results, you can make a statement about them. This statement communicates your results to others.

The Science Fair Project Presentation Board

The science fair project board is the visual representation of the students' hard work. Below is a list of the information that needs to be included along with where it is typically found on the project board. The students can certainly mix things up a bit, but be sure to remind them that their information needs to be placed in such a way that it is easy for someone else to follow.

The left section of the board typically has:

- ✓ Introduction
- ✓ Hypothesis
- ✓ Research

The center section of the board typically has:

- ✓ Materials
- ✓ Procedure
- ✓ Pictures, Graphs, and Charts from the Experiment (***Note****—The students can also display a portion of their project or a photo album with pictures from their experiment on the table in front of their board.*)

The right section of the board typically has:

- ✓ Results
- ✓ Conclusions

When purchasing a presentation board, you are looking for a tri-fold board that is at least 36" high and 48" wide, with the two side sections folding into the center section. If you search the internet for science fair project boards, you will find plenty of options, some with header boards, some without. You can usually purchase a project board at most local WalMart, Target, or Michael's stores.

Here is a description of what the students need to prepare for their presentation boards.

- **The Introduction** – Have the students turn their questions from step one into a statement. Then, they should write two to three more sentences explaining why they chose their specific topics. The students should end their introductory paragraphs by sharing the questions that they were trying to answer with their projects.
- **The Hypothesis** – Have the students type up and prepare their hypotheses from step three for the project boards.
- **The Research** – Have the students type up and prepare their research reports from step two for the project boards.
- **The Materials** – Have the students type up a list of the materials they used for their projects.
- **The Procedure** – Have the students revise the experiment designs they wrote in step four so that they are written in the past tense.
- **The Results** – Have the students turn the trends in the observations they noted and the results they interpreted in step six into a paragraph.
- **The Conclusion** – Have the students type up and prepare their concluding paragraphs from step six for the project boards.

Reading Assignments for Younger Students

Basics of Physics - Forces, Motion, and Energy

Unit 1 Motion

Week	Topic Studied	Resource and Pages Assigned
Week 1	Forces	*Usborne Science Encyclopedia pp. 118-121 (Forces)*
Week 2	Friction and Gravity	*Usborne Science Encyclopedia pp. 124-125 (Friction), 130-131 (Gravity)*
Week 3	Motion	*Usborne Science Encyclopedia pp. 122-123 (Dynamics)*
Week 4	Speed and Acceleration	*Usborne Science Encyclopedia pp. 126-127 (Motion)*

Unit 2 Energy

Week	Topic Studied	Resource and Pages Assigned
Week 5	Energy and Work	*Usborne Science Encyclopedia pg. 106 (Energy)*
Week 6	Energy Sources	*Usborne Science Encyclopedia pp. 108-109 (Energy Resources and Efficiency)*
Week 7	Pressure	*Usborne Science Encyclopedia pp. 132-133 (Pressure)*
Week 8	Simple Machines	*Usborne Science Encyclopedia pp. 134-135 (Simple Machines*

Concepts in Physics - Heat, Light, and Sound

Unit 3 Thermodynamics

Week	Topic Studied	Resource and Pages Assigned
Week 9	Energy Conversion	*Usborne Science Encyclopedia pg. 107 (Energy Conversion)*
Week 10	Heat	*Usborne Science Encyclopedia pp. 110-111 (Heat)*
Week 11	Thermodynamics	*Usborne Science Encyclopedia pp. 112-113 (Heat Transfer)*
Week 12	Engines	*Usborne Science Encyclopedia pp. 146-147 (Engines)*

Reading Assignments for Younger Students

Unit 4 Sound

Week	Topic Studied	Resource and Pages Assigned
Week 13	Sound	*Usborne Science Encyclopedia pp. 206-207 (Sound)*
Week 14	Sound Waves	*Usborne Science Encyclopedia pp. 202-203 (Waves)*
Week 15	Hearing Sound	*Usborne Science Encyclopedia pp. 372-373 (Ears)*
Week 16	Acoustics	*Usborne Science Encyclopedia pp. 208-209 (Musical Instruments)*

Unit 5 Light

Week	Topic Studied	Resource and Pages Assigned
Week 17	Light	*Usborne Science Encyclopedia pp. 212-213 (Electromagnetic Wave), pp. 214-215 (Light and Shadow)*
Week 18	Reflection and Refraction	*Usborne Science Encyclopedia pp. 218-219 (Light Behavior)*
Week 19	Vision and Color	*Usborne Science Encyclopedia pp. 216-217 (Color)*
Week 20	Optics	*Usborne Science Encyclopedia pp. 220-221 (Lens and Mirrors)*

Applications in Physics - Electricity, Magnetism, and Engineering

Unit 6 Electricity and Magnetism

Week	Topic Studied	Resource and Pages Assigned
Week 21	Electrical Current	*Usborne Science Encyclopedia pg. 229 (Static Electricity and Lightning)*
Week 22	Conductors and Insulators	*Usborne Science Encyclopedia pg. 228 (Electric Charge and Current)*
Week 23	Batteries	*Usborne Science Encyclopedia pg. 231 (Batteries*
Week 24	Circuits	*Usborne Science Encyclopedia pg. 230 (Circuits)*
Week 25	Magnetism	*Usborne Science Encyclopedia pp. 232-233 (Magnetism, except section on electromagnetism)*

Reading Assignments for Younger Students

Unit 6 Electricity and Magnetism, continued

Week	Topic Studied	Resource and Pages Assigned
Week 26	Electromagnetism	*Usborne Science Encyclopedia pp. 233-234 (Magnetism, begin with section on electromagnetism)*
Week 27	Motors and Generators	*Usborne Science Encyclopedia pg. 235 (Motors and Generating Electricity)*

Unit 7 Engineering and Robotics

Week	Topic Studied	Resource and Pages Assigned
Week 28	Engineering	*There are no pages for these topics in the Usborne Science Encyclopedia. Younger students should be able to handle the Bridges and Tunnels and Robotics books, as long as you read the pages out loud to them.*
Week 29	Bridges	
Week 30	Tunnels	
Week 31	Robotics	
Week 32	Actuators and Effectors	
Week 33	Sensors and Controllers	

Unit 8 Nuclear Physics

Week	Topic Studied	Resource and Pages Assigned
Week 34	Radioactivity	*Usborne Science Encyclopedia pp. 114-115 (Radioactivity)*
Week 35	Nuclear Energy	*Usborne Science Encyclopedia pp. 116-117 (Nuclear Power)*

Resultant Force Worksheet

Introduction

Objects have more than one force acting on them at any given time. If the forces are in the same direction, they add together. The effect of this addition on the object would be to accelerate, or move, it in that direction. If the forces are in opposing directions, they subtract or cancel each other out. The effect on the object depends upon the size of the opposing force. If the two forces are equal, they will balance each other out and the object will remain still. If one of the opposing forces is greater, the end result will be for the object to accelerate, or move, in that direction.

We can determine how an object will move by calculating the resultant force. The resultant force, which is also known as the net force, is the overall force acting on an object after all the forces are combined. To calculate the resultant force we use vector quantities to represent the forces. These vectors have both direction and size.

Sample Problems

Here are two sample problems for calculating the resultant force using vectors:

Resultant Force = 5N

The object will continue in the same direction.

Resultant Force = -4N

The object will begin moving in the opposite direction.

Problems

1.

Resultant Force = _______

The object will ____________________

____

2.

Resultant Force = _______

The object will ____________________

____

3.

Resultant Force = _______

The object will ____________________

____

4.

Resultant Force = _______

The object will ____________________

____

Second Law of Motion Worksheet

Introduction

Newton's second law of motion states that:

The greater the force on an object, the greater the change in its motion. The greater the mass of an object, the greater the force needed to change its motion.

This law can also be written as an equation which looks like:

Force (F) = mass (m) • acceleration (a)

We can use this equation to calculate the acceleration of an object when we know its mass and the force that has acted on it.

Sample Problems

Let's say you are pushing a grocery cart full of oranges that has a mass of 35 kg with a net force of 70 N. What would the acceleration of the cart be?

F = 70 N (kg • m/s^2)   m = 35 kg   a = ?

Substitute and solve.

70 N = (35 kg) a

$$a = \frac{70 \text{ N (or kg} \cdot \text{m/s}^2)}{35 \text{ kg}}$$

$a = 2 \text{ m/s}^2$

Problems

Now it is your turn to try a few problems!

1. A girl kicks a 2.5 kg soccer ball with a net force of 20 N. What would the acceleration of the ball be?

2. A truck pulls a trailer with mass of 4000 kg up a hill at a rate of 40 m/s^2. What is the net force that acted on the trailer?

3. A 12.5N force accelerates a boy in a toy car at 0.5 m/s^2. What is the mass of the boy?

Acceleration Worksheet

Introduction

Acceleration is the change in velocity over a period of time. We can calculate the acceleration of an object using the following formula:

$$\text{Acceleration (a)} = \frac{\text{change in velocity } (v_f - v_i)}{\text{total time (t)}}$$

If we have a positive acceleration, it means that the object is speeding up. If we have a negative acceleration, the object is slowing down.

Sample Problems

So, let's say we have a toy car that is stopped at the top of a ramp, which means its velocity is 0 m/s. The car is pushed down the ramp. After 3 seconds, the velocity of the toy car is 30 m/s. What is the car's acceleration?

$v_f = 30$ m/s $\quad v_i = 0$ m/s $\quad t = 3$ s $\quad a = ?$

Substitute and solve.

$$a = \frac{(30 \text{ m/s} - 0 \text{ m/s})}{3 \text{ s}}$$

$$a = \frac{30 \text{ m/s}}{3 \text{ s}}$$

$$a = 10 \text{ m/s}^2$$

Problems

Now it is your turn to try a few problems!

1. A student is sitting quietly enjoying the park, when suddenly he is scared by a spiny tailed lizard! He jumps off the park bench on which he is sitting and takes off running. He reaches the gas station down the road in 7 seconds. At that point he is moving 10 m/s. What is the student's acceleration?

2. A vehicle is driving along the road at 25 m/s. It hits a patch of ice and its speed increases to 35 m/s in 2 seconds. What is the vehicle's acceleration?

3. A female cheetah is chasing a gazelle at a speed of 28 m/s. She misses and ends the chase. Over the next 5 seconds, she slows down to a restful pace of 2 m/s. What is the cheetah's acceleration?

Work Worksheet

Introduction

Work is the transfer of energy that occurs when an object is moved or changed. In other words, work is the result of force applied over a distance. We use the following equation for calculating the amount of work in a giving situation:

Work (W) = Force (F) • Distance (d)

The unit for work is the Joule (J), which stands for newton-meter (N•m). Now that we understand what work is and how to calculate it, let's look at a sample problem.

Sample Problems

A man is pushing a block with a 5 N force. He travels 15 meters. How much work did he do?

Force (F) = 5 N   Distance (d) = 15 m   Work (W) = ?

W = 5 • 15

W = 75 J

The man has done 75 J of work.

Problems

Now it is your turn to try a few problems!

1. A man used a 4 N force to push his stalled car the remaining 50 m to his house. How much work has the man done?

2. A person carries a bag of dog food that weighs 20 N up 12 m of stairs. How much work has that person done?

3. Once it has been filled, a hot air balloon weighs 900N. It quickly rises 700m. How much work did the hot air do?

4. A mechanical winch is attached to a bundle of lumber weighing 250 N. When you turn the winch on and use it to lift the lumber 4.2 m to the second floor window, how much work has the winch done?

Pressure Worksheet

Introduction

A sharp pencil point is easier to push through paper than a blunt eraser. This is because the pencil point has a smaller area than the eraser. So, when pushed with the same force, the point exerts greater pressure than the eraser.

Pressure is the result of force distributed over area. We use the following equation to calculate force:

$$\text{Pressure (P)} = \frac{\text{Force (F)}}{\text{Area (a)}}$$

The unit for Pressure is Pascal (Pa) which is equal to Newton per meter squared (N/m^2)

Sample Problems

What is the pressure exerted by 5000 N of water over an area of 25 m^2?

Force = 5000 N   area = 25 m^2 P = ?

Substitute and solve.

$$P = \frac{5000}{25}$$

P = 200 Pa

Problems

Now it is your turn to try a few problems!

1. A man hammers a nail in with a force 250 N. The nail point is 0.10 m^2. What is the pressure exerted by the nail as it makes its way through the wood?

2. A latex balloon is blow up to have an area of 1.6 m^2. The hot air inside pushes with a force of 100 N. How much pressure is being exerted on the latex?

3. A box weighing 2400 N is sitting on the floor. If the area of the box touching the ground is 1.5 m^2, what is the pressure being exerted on the floor by the box?

Converting Temperatures Worksheet

Introduction

There are three major scales used for reporting a temperature - Kelvin, Fahrenheit, and Celsius. The Kelvin scale is used mainly in scientific applications. The Fahrenheit and Celsius scales are used more in everyday applications, so it is important to know how to convert between them. When converting between these scales, we use the following formula:

$$^{o}F = 1.8 \cdot {^{o}C} + 32$$

^{o}F stands for the temperature in Fahrenheit. ^{o}C stands for the temperature in Celsius.

Sample Problems

What is the temperature of the Sun in ^{o}F? (The temperature of the Sun is 5505 ^{o}C.)

$^{o}C = 5505$ $^{o}F = ?$

Substitute and solve.

$^{o}F = 1.8 \cdot (5505) + 32$

$^{o}F = 9909 + 32$

$^{o}F = 9941$

The temperature of the Sun is 9941 ^{o}F.

Problems

Now it is your turn to try a few problems!

1. Convert 32 ^{o}C into ^{o}F.

2. Convert 16 ^{o}C into ^{o}F.

3. Convert 56 ^{o}C into ^{o}F.

4. Convert 300 ^{o}F into ^{o}C.

5. Convert 32 ^{o}F into ^{o}C.

Specific Heat Worksheet

Introduction

The amount of heat energy needed to change the temperature of a substance depends upon what the substance is, how much of it you have, and the rise in the temperature. So, in other words, the change in the temperature of your car's hood will be higher than that of the plastic light covers because the metal in the hood is more likely to absorb heat energy.

We can calculate this ability using the material's specific heat capacity, which is the amount of energy required to raise the temperature of one gram of substance by one degree Celsius. To calculate the heat absorbed by a given material, we use the following equation:

$$Q = m \bullet c \bullet \Delta T$$

The Q stands for the heat that is absorbed by a material, which is measured in Joules (J). The m stands for mass, which is measured in grams (g). The c stands for specific heat capacity (J/g•°C) and the ΔT stands for the change in temperature using the Celsius (°C) scale.

Sample Problems

An iron manhole cover has a mass of 600 grams. How much heat must be absorbed to raise the manhole cover's temperature by 15 °C? The specific heat of iron is 0.449 J/g•°C.

m = 600 g   c = 0.449 J/g•°C ΔT = 15 °C

Substitute and solve.

$$Q = (600)(0.449)(15)$$

$$Q = 4041 \text{ J}$$

Problems

Now it is your turn to try a few problems!

1. How much heat is needed to raise the temperature of 375 g of water by 25 °C? The specific heat of water is 4.18 J/g•°C.

2. How much heat is absorbed by a 575 g copper skillet when its temperature rises from 22°C to 147°C? The specific heat of copper is 0.385 J/g•°C.

3. To release an amethyst from its setting, a jeweler heats a 12 g silver ring by adding 26.4 J of heat. How much does the temperature of the silver increase? The specific heat of silver is 0.235 J/g•°C.

Power Worksheet

Introduction

An engine's ability to move a vehicle depends upon how powerful it is. Power measures the rate at which the engine does work. We use the following equation to calculate the power needed:

$$P = \frac{W}{t}$$

The unit for power is Watts (J/s). (**Note—***Often, in a power problem, we are given the force and the distance an object is moved over a given time. Remember that work is equal to the force times the distance*.)

Sample Problems

You exert a force of 67 N to lift a stack of books to a height of 0.7 m in a time of 3 seconds. How much power did you use to lift the books?

$W = F \cdot d = (67)(0.7) = 47\text{ J}$ $\quad t = 3\text{ s}$ $\quad P = ?$

Substitute and solve.

$$P = \frac{47}{3}$$

$$P = 15.7\text{ Watts}$$

Problems

Now it is your turn to try a few problems!

1. An athlete rows a boat across a still pond, doing 5000 J of work in 40 seconds. What is the rower's power output?

2. A weightlifter raises a barbell 1 m with a force of 100 N in 1.3 seconds. How much power did he use to lift the weights?

3. A truck pulls a trailer at a constant velocity for 387 m while exerting a force of 520 N for 1.5 minute (90 seconds). Calculate the power.

Mechanical Wave Worksheet

Introduction

If you remember from the motion unit, the speed of an object equals distance divided by time. In the same way, we can measure the speed of a mechanical wave by taking its length and the time it takes to complete a period. The period of a wave is the time it takes to go from one crest of the wave to the next crest of the wave.

The period of a wave is very tiny, so we use the wave's frequency instead to calculate the speed of a wave. Frequency measures the number of periods (or vibrations) a wave completes in one second. So, to calculate the speed of a wave, we use the following equation:

Speed (v) = Wavelength (λ) • Frequency (f)

The SI unit of frequency is Hertz (Hz), which stands for one vibration per second.

Sample Problems

Let's say you have a string on a musical instrument, which you have plucked. The plucking produces a wave with the length of 0.28 m and frequency of 4 Hz. What would the speed of the wave be?

Speed (v) = ?   Wavelength (λ) = 0.28 m   Frequency (f) = 4 Hz

Substitute and solve.

$v = (0.28 \text{ m})(4 \text{ Hz})$

$v = 1.12 \text{ m/s}$

Problems

Now it is your turn to try a few problems!

1. A rope is vibrated, producing a wavelength of 3m. The frequency of the wave is 0.7 Hz. What is the speed of the wave?

2. What is the speed of a wave in a spring that has a wavelength of 0.2 m and frequency of 5 Hz?

3. A sound wave at sea level has a wavelength of 6 m. What is its frequency? The speed of sound at sea level is 340.29 m/s.

4. What is the wavelength of an earthquake wave that travels at a speed of 3km/s and has a frequency of 8 Hz?

Electromagnetic Wave Worksheet

Introduction

When we looked at mechanical waves, we used the wave speed equation to calculate their speed or frequency. However, electromagnetic waves all travel at the same speed in a vacuum. That speed is the speed of light, which is 3.0×10^8 m/s. Electromagnetic waves do vary in wavelength and frequency, so we can use the wave speed equation to calculate those. As a refresher, here is the wave speed equation:

Speed (v) = Wavelength (λ) • Frequency (f)

The SI unit of frequency is Hertz (Hz), with stand for one vibration per second.

Sample Problems

So let's say you know that a radio station broadcasts an electromagnetic radio wave with a wavelength of 2.6 meters. What is the frequency of the wave?

Speed (v) = 3.0×10^8 m/s Wavelength (λ) = 2.6 m Frequency (f) = ?

Substitute and solve.

$$3.0 \times 10^8 = (2.6) f$$

$$f = \frac{3.0 \times 10^8}{2.6}$$

$$f = 1.2 \times 10^8 \text{ Hz}$$

Problems

Now it is your turn to try a few problems!

1. A cable station satellite sends an electromagnetic wave with a wavelength of 3 m. What is the frequency of the wave?

2. What is the frequency of an electromagnetic wave that has a wavelength of 6 m?

3. What is the wavelength of an electromagnetic wave that has a frequency of 1.9×10^7 Hz?

Electrical Charge Worksheet

Introduction

When current flows for a set amount of time, we can calculate the charge. This charge is reported in Coulombs (C) – 1 Coulombs (C) is equal to 6.242×10^{18} electrons). The equation to calculate electrical charge is:

$$Q = I \bullet t$$

The Q stands for electrical charge, which is measured in Coulombs. The I stands for current, which is measure in amperes, and t stands for time measured in seconds.

Sample Problems

So, if there is a current of 26 amperes in a circuit for 3 minutes, what quantity of electric charge flows through the circuit?

I = 26 amps t = (3 min)(60 sec/min) = 180 sec Q = ?

Substitute and solve.

$$Q = (26)(180)$$

$$Q = 4680 \text{ C}$$

Problems

Now it is your turn to try a few problems!

1. If there is a current of 45 amperes in a circuit for 2 minutes, what quantity of electric charge flows through the circuit?

2. If there is a current of 10 amperes in a circuit for 1 hour, what quantity of electric charge flows through the circuit?

3. How much current must there be in a circuit if 476 coulombs flows past a point in the circuit in 7 seconds?

4. How much time is required for 43 coulombs of charge to flow past a point if the rate of flow is 3.2 amperes?

Potential Difference Worksheet

Introduction

Potential difference is the amount of work done by a charge that is moving between two points, divided by the size of the charge. In simpler terms, it measures the likelihood of a charge moving from one place to another. The greater the potential difference, the more likely that the charge will move. We use the following equation to calculate the potential difference:

$$V = \frac{E}{C}$$

In this equation, V stands for potential difference, which is also known as the voltage. The E stands for the energy transferred in Joules and C stands for the electrical charge, which is measured in Coulombs.

Sample Problems

A battery uses 90 J of energy to transfer 10 C of electrical charge between the two terminals. What is the voltage of the batter?

$E = 90$ J $\quad\quad$ $C = 10$ C $\quad\quad$ $V = ?$

Substitute and solve.

$$V = \frac{(90)}{(10)}$$

$V = 9$ volts

Problems

Now it is your turn to try a few problems!

1. You need 45 Joules of work energy to transfer 3 C of electrical charge between two points. What is the potential difference between the two points?

2. A circuit requires 600 J of energy to transfer an electrical charge between two points in the circuit, which have a potential difference of 30 volts. How much charge is transferred?

3. A 6 volt battery transfers 13 C of electrical charge between the two terminals. How much energy is transferred?

Ohm's Law Worksheet

Introduction

Georg Ohm, a German physicist, found that if the voltage in a circuit increases, so did current. He also found that if the resistance in a circuit increased, the current decreased. He developed a mathematic explanation for what he saw, which we know as Ohm's Law. Ohm's Law can be written as the following equation:

$$V = I \cdot R$$

In Ohm's Law, V stands for voltage, the I stands for current, and the R stands for resistance, which is measured in ohms.

Sample Problems

A light bulb has a resistance of 3.5 ohms and a current of 1.7 amps. What is the voltage across the bulb?

$I = 1.7$ amps $\quad R = 3.5$ ohms $\quad V = ?$

Substitute and solve.

$V = (1.7)(3.5)$

$V = 5.95$ volts

Problems

Now it is your turn to try a few problems!

1. What is the voltage of a circuit with 12 amps of current and 3 ohms of resistance?
2. If a toaster produces 10 ohms of resistance in a 140-volt circuit, what is the amount of current in the circuit?
3. A circuit contains a 9 volt battery and light bulb with a resistance of 3 ohms. What is the current?
4. How much voltage would be necessary to generate 15 amps of current in a circuit that has 6 ohms of resistance?

Templates

Two Days a Week Schedule Template

Week: ____________________ Topic: ____________________

Day 1	Day 2
Supplies I Need for the Week	
Things I Need to Prepare	

Week: ____________________ Topic: ____________________

Day 1	Day 2
Supplies I Need for the Week	
Things I Need to Prepare	

Five Days a Week Schedule Template

Week: ____________________ Topic: ____________________

Day 1	Day 2	Day 3	Day 4	Day 5

Supplies I Need for the Week

Things I Need to Prepare

Week: ____________________ Topic: ____________________

Day 1	Day 2	Day 3	Day 4	Day 5

Supplies I Need for the Week

Things I Need to Prepare

Scientist Biography Report Grading Rubric

Spelling (points x 1)
- ✓ 4 points: No spelling mistakes.
- ✓ 3 points: 1-2 spelling mistakes and not distracting to the reader.
- ✓ 2 points: 3-4 spelling mistakes and somewhat distracting.
- ✓ 1 point: 5 spelling mistakes and somewhat distracting.
- ✓ 0 points: > 5 spelling mistakes and no proofreading obvious.

Points Earned __________

Grammar (points x 1)
- ✓ 4 points: No grammatical mistakes.
- ✓ 3 points: 1-2 grammatical mistakes and not distracting to the reader.
- ✓ 2 points: 3-4 grammatical mistakes and somewhat distracting.
- ✓ 1 point: 5 grammatical mistakes and somewhat distracting.
- ✓ 0 points: > 5 grammatical mistakes and no proofreading obvious.

Points Earned __________

Introduction to the Scientist (points x 2)
- ✓ 4 points: Includes thorough summary of the scientist's biographical information and why the student chose the particular scientist.
- ✓ 3 points: Adequate summary of the scientist's biographical information and why the student chose the particular scientist.
- ✓ 2 points: Inaccurate or incomplete summary of one of the scientist's biographical information and why the student chose the particular scientist.
- ✓ 1 point: Inaccurate or incomplete summary of both of the scientist's biographical information and why the student chose the particular scientist.
- ✓ 0 points: No introduction

Points Earned __________

Description of the Scientist's Education (points x 2)
- ✓ 4 points: Includes thorough summary of the scientist's education.
- ✓ 3 points: Adequate summary of the scientist's education.
- ✓ 2 points: Inaccurate or incomplete summary of one of the scientist's education.
- ✓ 1 point: Inaccurate or incomplete summary of both of the scientist's education.
- ✓ 0 points: No description of the scientist's education.

Points Earned __________

Description of the Scientist's Major Contributions (points x 2)
- ✓ 4 points: Includes thorough summary of the scientist's major contributions.
- ✓ 3 points: Adequate summary of the scientist's major contributions.
- ✓ 2 points: Inaccurate or incomplete summary of the scientist's major contributions.
- ✓ 1 point: Inaccurate and incomplete summary of the scientist's major contributions.

- ✓ 0 points: No description of the Scientist's Major Contributions and Interesting Facts of their life.

Points Earned __________

Conclusion (points x 2)

- ✓ 4 points: Explanation of why the student feels one should study the scientist and a summary statement about the scientist.
- ✓ 3 points: Adequate explanation of why the student feels one should study the scientist and a summary statement about the scientist.
- ✓ 2 points: Incomplete or incorrect explanation of why the student feels one should study the scientist and a summary statement about the scientist.
- ✓ 1 point: Conclusion does not have an explanation of why the student feels one should study the scientist and a summary statement about the scientist.
- ✓ 0 points: No conclusion.

Points Earned __________

Final Score = (Total Points/40) x 100%

Total Points Earned _______________

Final Score _______________

Salt Dough Recipe

Ingredients

- ✓ 1 cup (273 g) salt
- ✓ 2 cups (256 g) flour
- ✓ ¾ cup (180 mL) warm water

Directions

1. In a large bowl mix the salt and flour together.
2. Gradually stir in warm water and mix until it forms a doughy consistency.
3. Turn the salt dough onto the bench and kneed with your hands until smooth and combined.

91471516R00148

Made in the USA
Columbia, SC
22 March 2018